how to b
wind pu

Jim Barr

Leaves rustle,
Blades turn,
Water moves.

To Ian Harrison
who still understands the wind

L I L I

Published February 2006 by

Low-Impact Living Initiative
Redfield Community, Winslow, Bucks, MK18 3LZ, UK
+44 (0)1296 714184

lili@lowimpact.org
www.lowimpact.org

Printed in Great Britain by
Lightning Source, Milton Keynes

ISBN 0-9549171-2-X

contents

table of figures

about the author

Jim Barr was born in 1944, and now lives in Bedfordshire, England with his wife and two sons. He started his career in research at Imperial College London as a technician in physics, and has since worked as a jackaroo on an Australian cattle station, a deck hand on a French fishing boat, a developer and lecturer in computer software. He has designed microwave detection equipment for the nuclear industry, and written on adventure and renewable energy.

If you wish to contact me, you can email me at wandana@ntlworld.com, or write to:

Machine Conversation
34 The Ridgeway,
Flitwick,
Bedfordshire, MK45 1DH
UK

In spite of the effort of several proof readers it is possible that some errors remain; it is important that you bear this fact in mind when implementing the design.

I will be grateful for comments and constructive criticism from readers so that I may make modifications to any later editions of the book.

disclaimer

It is up to you to check that your site is suitable for this kind of wind pump, and if building your own, that your workshop, tools, and skills are up to the job. Anyone using the methods described in this manual does so at their own risk. LILI, the author, and their associates assume no responsibility for damage to persons or property caused by the building, installation or use of a wind pump using this book as a guide.

glossary

Annulus (see DIY foot valve, p.76): the area between two concentric circles, or the area of a ring.

Asterisk *, when used in technical drawings, means times (x) as in multiply (arithmetic). So for you non-engineers, 3 * 4 means 3 x 4.

Cad pas (steel) - see Figures 15 & 16: short for cadmium passivated; this is a treatment for steel components to prevent corrosion, and has a yellowish colour.

Duralumin (see connecting rod, p.47): a corrosion resistant alloy of aluminium, copper, manganese, magnesium, iron and silicon.

Furl (see p.83): to turn a turbine out of high winds to slow it down; to brake.

Oilite (see connecting rod, p.49): oilite bearings are self-lubricating because they are impregnated with oil which forms a film between the bearing and the shaft. This dramatically reduces friction.

Sintered bronze (see Figure 17): often used for bushes and bearings because its porosity allows lubricants to flow through it - made by pressing powdered bronze into the desired shape and heating to harden.

Swing over bed (when talking about lathe work): the maximum diameter of the work that can be turned in the lathe.

Yaw (see p.85): the yaw mechanism of a horizontal-axis wind turbine turns it into the wind, so that it can extract the maximum amount of energy.

preface

This book follows on from a book I wrote in 1994 called 'Building a Domestic Wind Pump'. There are references in the text of this book to the first one, and you may build the first design using this new book without needing the previous one.

Several minor errors in the first book have been rectified. There are several other changes to be found in this new book; these resulted from the ongoing development work carried out on the early design. The only change that causes me concern is the inclusion of a commercial foot valve. This contradicts my aim of creating a design that may be built entirely from readily available and, where possible, scrap materials; but the home-made foot valve described in the first book, although quite reliable, was sometimes a bit unpredictable and difficult to maintain. I have left the original design in the this book because you may decide to try one anyway.

In many cases I have made reference to the original design, in areas where the new design is quite distinct. I hope this will not confuse; my purpose is to give the reader the opportunity to follow the same evolutionary route that I took. In many cases you may leave my own plan behind and arrive at a different design, driven by resource limitations, availabilities or just a better imagination. For many, this book will be merely a catalyst.

While my experience of wind pumps has been gained in a variety of environments and climates, this book was originally written to reflect my perception of domestic wind pump requirements in Britain, to move domestic grey water for irrigation or some sort of re-use rather than put it down the drain. Many copies of my first book were sold overseas however, and the response and comments that I received led me to re-examine the limited approach of the first book; it is apparent that many people are looking to small wind pumps for the solution of several non-domestic water problems.

I have continued to experiment with many of the ideas that have been prompted by the first book, and this book contains all of the improvements that are current at the time of writing.

May I thank all the people who, having read and used the first book, have made constructive criticisms, which have led I hope, to improvements in this book.

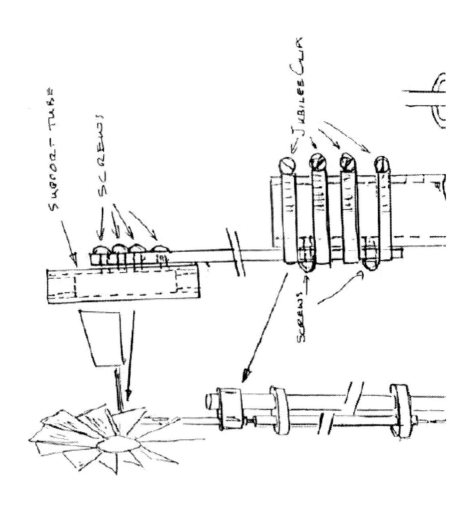

SUPPORT TUBE

SCREWS

JUBILEE CLIP

SCREWS

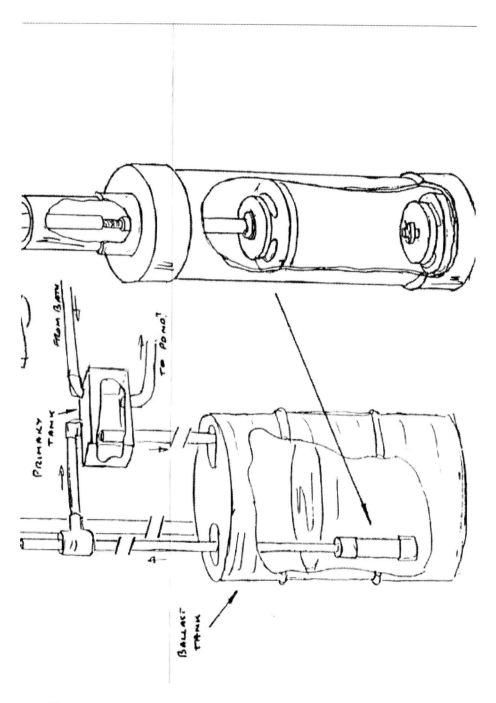

Figure 1: preliminary drawing by the author

introduction

The wind has been utilised by man for 6000-7000 years, starting with sailing boats. Land-based wind machines first appeared around 300-500 AD. Most of the designs we see around today are based on ideas from 1200 AD onwards.

My own interest in wind power was inspired in the early 70s during my travels in the Australian bush. If I was short of water for the engine or for myself, I would look for a turbine in the distance and head for it. The water from such pumps was for stock and tasted foul but in desperation it was drinkable.

Still later, back in the UK in the mid 80s, during the 'drought', I got fed up with carrying bath water up the garden to put on our trees and vegetables, and thus the chain of development started that resulted in my current system and this book.

As I complete this book (autumn 2005), we are still experiencing very confusing weather, but the trend is warmer and in the UK drier. In any event, the unpredictable nature of the climate means that the need for wind pumps of the type described in this book will increase considerably.

water extraction and the law

I do not know about the law regarding the extraction of water, but it's a safe bet that you can't do it without a licence of some sort. I live in Bedfordshire, England and have been verbally informed by Anglian Water that it is OK to divert the domestic grey water to the garden by any means I choose. If I can prove that I am reducing the load on the sewerage system, I can get a rebate on the annual charges I pay. You should consult your local water board or company.

While it is OK in my area to redirect grey water to the garden, the same is not true about redirecting surface (rain) water to the soil drain; so if you are collecting water from a variety of sources to be recycled, and need an overflow to accommodate days when the wind is not blowing, you may be in trouble if part of that water is rain water and you divert it to the soil drain. My own water board permit me to do this, but I believe that this is a special case, too complicated to go into here. Email me if you want to know more!

planning permission

This will always be a local issue and you must contact your local planning authority. In urban areas in the UK, you will usually be able to raise one of my turbines to 5 metres without a problem; above 7 metres it may be a problem with neighbours, even if planning permission is granted.

the wind pump

what kind of wind pump?

The type of wind pump discussed in this book is known as a multi-blade direct drive wind pump. This description distinguishes it from wind-powered electric pumps, geared wind pumps, and several other designs most of which are usually more complex from the DIY engineer's point of view. In fact one of the more recent developments has been the inclusion of a chain-drive gear system which means that if you choose to add this, then it is no longer a direct drive pump. Both designs are included in this book.

This wind pump is not a toy but it is small; in a light-to-moderate wind, the system should pump about 1000 litres per day with a head of 2-3 metres.

Most wind pumps of this type function better when working against a larger head of water than that quoted above. This wind pump has evolved for use in the domestic environment and most of the applications I have come across do not require large heads of water to be moved.

The pump and turbine described here should work OK with increased heights, but you may encounter some mechanical problems when you try to extend the distance between the turbine and pump by too much. These problems result from my use of lightweight flexible materials for the transmission system.

why have a wind pump?

Even if you live in a semi-rural part of the UK, your reasons for choosing wind energy to pump domestic water are unlikely to be based on sound economics. There are, however, many other good reasons for using wind instead of a fossil fuel-based energy source for a water pump.

Long before it became fashionable to be green, I derived some pleasure from utilising this free, renewable energy source; by recycling domestic grey water to the garden, I was able to save our fruit trees and bushes during several dry summers.

Water is set to become a very scarce resource in the 21st Century. Any method of utilising what would otherwise be waste water has got to be a good idea.

To me, the wind pumps in my garden are attractive to look at even when stationary, and the pumps do useful work irrigating the garden. Electric pumps could do this but they would incur running costs and are not so attractive (to me).

As an engineer, I get pleasure from watching the simple mechanisms that convert energy into work; I imagine that some people go to traction engine rallies or build model steam locomotives for similar reasons.

For the amateur, there is very little invention left in wind pump design these days, since the basis of the majority of successful designs has been around for a long time (1600 years); but there are opportunities for the amateur / experimental engineer to make minor changes to established designs and observe the resulting changes in performance. I am currently working on two designs for wind pumps which bear no resemblance to any previous systems, but they are both rather theoretical and will require much more time and investment to be proved useful (or useless).

[I wrote the above paragraph in early 1998 and have decided that the 'theoretical' design I was working on is best kept that way - theoretical. I have ceased diverting funds in that particular direction, but I may still return!]

what can it do?

This system does NOT generate electricity!
However it can:

clean and allow reuse of grey water
I have included some detail about cleaning grey water with biological and mechanical filters.

pump collected rainwater to where it is needed
See 'resources' for rainwater harvesting links.

irrigate a vegetable plot
The simplest way to do this is to collect the pumped, cleaned water in a tank and transfer from the tank with a watering can, but you can also allow the tank to empty slowly via a hose pipe laid by the points to be irrigated, small perforations having been made in the pipe to allow a controlled leak onto the vegetables and flowers.

aerate a fish tank / pond
One of my systems was bought by a farmer solely for this purpose. He had many valuable fish in a large pond (6000 gallons), but did not want his new fish to go into the main pond until they had been quarantined for some weeks. It was simpler for him to have a separate smaller tank with water circulated by wind power than install more electric power in the required area.

be a bird scarer
Not the most efficient use of the technology, but I do have one client who was unable to place the turbine where he wanted it, and simply put it in a field doing nothing but turn in the breeze. He reckoned it was the best bird scarer for his crops ever!

supply water to small community in a developing country
A major opportunity; there are many developing societies that rely on low-tech solutions for pumping water from wells.

Sometimes the community is large enough to enable the economic use of a large wind turbine. For smaller communities this is not viable and a system like mine is the only option. I have to say that the smaller systems from Poldaw (see resources) are very much in this market, and although too large for a really small community, they do provide a superb system at a relatively low cost.

pump from well, bore, pond, river
This really goes without saying. The problems, if any, only arise if the water is at a site where the wind is not!

be part of a water feature in a garden
There are numerous examples in California where wind turbines are used as decoration; some of them do not even turn! To have a simple wind turbine in a garden just creating a water feature, fountain or 'burble' over stones can make a delightful addition to the ambience.

how does it work?

The wind causes the turbine rotor to turn (revolve); this turns a crank which converts the rotation into vertical oscillation in a transmission rod; this rod raises and lowers a piston in a pump comprising a cylinder and two valves. During the down stroke the cylinder fills with water, and during the up stroke the piston raises the water in the cylinder and riser.

who can build one?

This depends on how much you wish do yourself and how well you are equipped.

If you get all the parts manufactured, it is not much more complicated than DIY flat-pack furniture!

If you want do it all yourself, then you will need a fairly extensive workshop with skills and equipment for lathe turning, sheet metal cutting, punching and bending, and some welding.

If you have a good engineering mind, some imagination and some skills and equipment you can fabricate most of the components without welding or lathe turning. As I have said already, this book should be seen as not just informative, but hopefully, as a catalyst.

All components of the wind pump are described in some detail and the materials are part of that description; however, it may be useful to have an overview of the whole project to establish what technologies are required.

You will need to be able to accurately cut, bend, drill, rivet, turn, hard solder, soft solder, weld and / or glue a variety of materials. The processes, together with their levels of difficulty are summarised below.

	Aluminium	Brass	PVC	Wood	Steel
Line cut	3	3	0	0	2
Circle cut	3	3	3	0	2
Drill	4	4	3	3	2
Tap	4	3	5	0	4
Screw	0	3	2	0	3
Bend	6	0	0	0	6
Glue	0	0	2	0	0
Turn	0	5	3	3	0
Solder	0	7	0	0	0
Weld	0	0	0	0	6

The number in the chart indicates the degree of skill required for the associated process: e.g. 0 means you don't have to do it; 1 means very basic skill level; 9 means that you may need precision skills and / or possibly specialist tools. Sections in *italics* refer to the older design only.

As a guide I would say that, using this scale, building a typical tissue and balsa wood 24-inch wing-span model airplane from a kit rates as 6.

where to site a wind pump

There are three important considerations when selecting a site for a wind pump. The first relates to how near your neighbours are and whether they will object to a 12-30 foot tower overlooking their garden.

The other two problems to be considered relate to positioning the wind pump for optimum performance.

The wind turbine and hence the mast of a direct-drive or gear-drive wind pump needs to be directly above the pump to which it is attached. For this reason you will try to site your wind pump as near to the source of the water to be moved as possible. Some degree of latitude may be allowed here, especially if the water is being pumped from a storage tank; it may be possible to pipe the water some distance (downhill) from the tank to the foot of the wind pump (see fig 2).

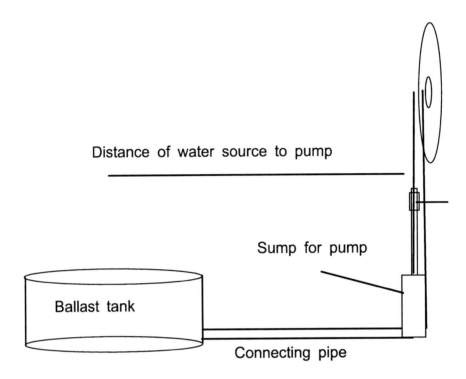

Figure 2: site the turbine above the pump

The turbine needs to be in clean (non-turbulent) air. For most domestic gardens this presents the main problem, since buildings, trees and other structures will disrupt the airflow to varying degrees.

Figure 3 [avoiding turbulence] shows the turbulent envelope to be avoided upwind and downwind of buildings and trees of height H. In an urban environment you are very unlikely to be able to site a wind pump outside this envelope - but it is not a simple on or off situation, and the wind pump I shall describe will operate in some pretty turbulent areas. One serious problem with a small wind turbine in turbulent air is the way in which the rotor can be buffeted to all points of the compass; this is not good for the mechanisms and is of course very inefficient.

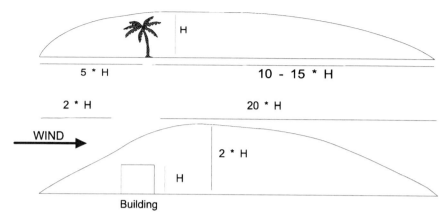

Figure 3: avoiding turbulence

The problem can be reduced by selecting a site for the wind pump after careful examination of the wind movements over the available ground over a period of time. A cheap comparative test for clean air flow is to stick several long poles in the ground at points of interest, each with a piece of light cloth or plastic tied to the top. By observing the wind's effect on the different flags, a picture of the air flow will develop. It is important to differentiate between rapid random turbulent agitation (no good) and a steady, though less aggressive flow (better).

Since the wind direction changes seasonally in most areas of the world, these tests should be carried out during the seasons for which pumping is required.

how to approach the building of a wind pump

As I said earlier, for many people, this text will be no more than a design catalyst to prompt your own methods of construction. If however, you are relying on the detailed descriptions within this book to help you build your first wind pump, then read all the relevant sections to get a clearer picture of how to do it.

Some parts of the wind pump described in this book are covered in more than one section - for example, the foot valve is part of the pump body and is also described in detail under the heading of valves.

This book includes descriptions of most of the incremental design changes that have occurred during the evolution of the system; so it is necessary to read all the detail to put each development in perspective.

In fact most of the design changes for the components are compatible with their previous assembly counterparts, so provided you are prepared to check and are confident of your own skills, you can 'mix & match' old with new.

Most of the components described in this book are over-engineered and some latitude may be exercised in implementing the design. I hope that common sense will decide where, and to what extent it is OK to depart from the suggestions in the book.

Just remember that the turbine weighs several pounds, it moves quite fast, has sharp edges and has a fair way to fall if bits of it break off.

health & safety

If you already have extensive experience of DIY, using a wide range of power and bench tools then you are probably aware that even small-scale engineering is potentially dangerous and costly when things go wrong. If, however, you do not feel that you have much experience in the tasks outlined in this book, I recommend that you pause to consider whether you should spend money on getting someone else to make the bits and confine yourself to assembly. I am not trying to be patronising but I have heard tales of good smart folk finding out the hard way that mistakes can hurt.

When I worked as a technician with Rank Xerox in Australia, they had a card that said something like:

'Field trials have shown that sprockets, chains and fast-moving metal parts are stronger than fingers.'

You should also be aware that if something you build causes injury to a third party, you may be sued.

cost of building a wind pump

My first wind pump (1989) was 1 metre diameter; it was built entirely from scrap and was almost useless - no, it *was* useless: it barely illustrated the principles on which wind pumps work. Given a hacksaw and a few other basic hand tools, you could build one like it for under £10. You could choose to do it this way, or you can buy a commercial machine for between £1600 and £16,000. I have not seen any commercial system smaller than 1.8 metre diameter. The third option is to learn from my mistakes and build a basic 12-blade wind pump as described in this book. If you have to buy everything from ironmongers and get a few parts machined or pressed, it could cost as much as £500, but if you have access to a reasonable workshop and a metal scrap bin then the cost could be under £100.

equipment: hand tools

A normal set of workshop hand tools for metal working is essential including hammer, hacksaw, files, hand drill, good pop-riveter etc.

equipment: power tools

A small pillar drill will make some operations simpler but it is not essential. A lathe capable of turning 20cm between centres and a swing over bed of 10cm is probably essential if you intend making everything yourself, but only a very few components rely on such equipment. In the later designs I have welded steel components rather than use hard solder.

other tools

A fly-press or a device for bending sheet metal would be useful, but not essential. A lot of the repetitive tasks are rendered easier if hard wood or 'dural' jigs are made to ensure consistency. This is especially true of the turbine blades; if you do not wish to make a jig to help bend the blades, then I still recommend the use of patterns to help mark them out: one to check that each blade is the same and one to help drill all the holes in the same place.

building the wind pump

turbine components and assembly

Figure 4 [exploded view of original main turbine components] gives an overall idea of what the components look like together. This is a view of the original design and is NOT exactly what the system will look like if you use the components for the more recent design.

Figure 5 [exploded view of intermediate main turbine components] shows how the intermediate model should look - updated but still direct-drive (without gears).

Figure 6 [exploded view of latest (geared) main turbine components] Is my final (current) design showing how the geared model should look.

In Figure 6, the fin pole bracket is shown with a twisted end. There is an even later design for the fin pole bracket - shown in Figures 37 and 38 - in which the bracket is also twisted but in a more complex manner.

You do NOT need the final swinging fin pole design if you are in the position to manually furl the rotor during high winds.

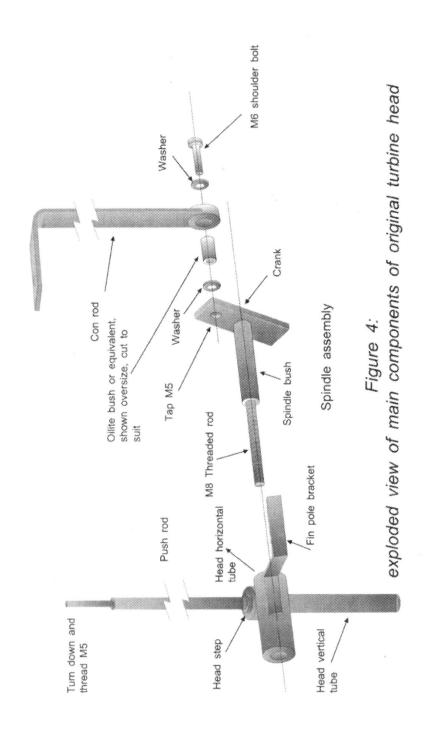

Turn down and thread M5

Push rod

Head horizontal tube

M8 Threaded rod

Head step

Head vertical tube

Con rod

Oilite bush or equivalent, shown oversize, cut to suit

Tap M5

Washer

Washer

M6 shoulder bolt

Crank

Spindle bush

Fin pole bracket

Spindle assembly

Figure 4:
exploded view of main components of original turbine head

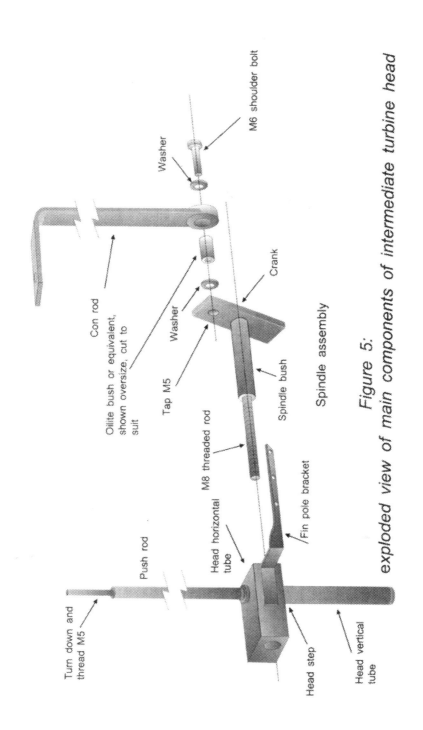

Turn down and thread M5

Push rod

Head horizontal tube

M8 threaded rod

Head step

Head vertical tube

Fin pole bracket

Spindle bush

Crank

Spindle assembly

Con rod

Oilite bush or equivalent, shown oversize, cut to suit

Tap M5

Washer

Washer

M6 shoulder bolt

Figure 5:
exploded view of main components of intermediate turbine head

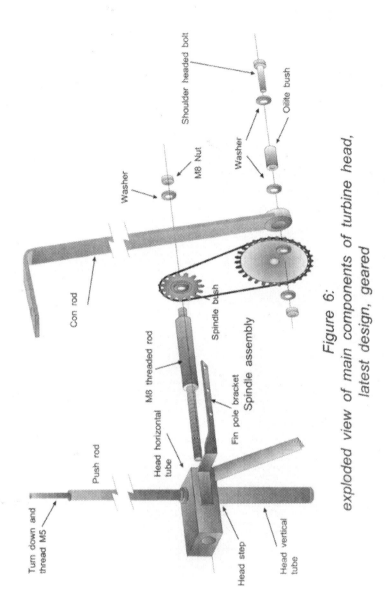

Figure 6:
exploded view of main components of turbine head,
latest design, geared

Shoulder headed bolt

Oilite bush

Washer

Washer

M8 Nut

Con rod

Spindle bush

M8 threaded rod

Spindle assembly

Fin pole bracket

Head horizontal tube

Push rod

Turn down and thread M5

Head step

Head vertical tube

These figures include most of the components above the support tube, (which is attached to the mast head). You may need to refer back to Figures 4, 5 and 6 when looking at the detailed descriptions given later.

The whole turbine assembly comprises two groups of parts. Group one is the rotor itself. It has four basic components: a front hub plate; a back hub plate; a wooden spacer; and the 12 blades with 48 pop rivets. Group two consists of five parts: the spindle; the spindle bush; the crank; the connecting rod; and the push rod.

Because the whole turbine is so small (only 700mm across), it is possible to make the blades strong enough and heavy enough to be a structural part of the assembly. When the blades are pop-riveted to the front and back hub plates, the resulting structure is very solid indeed.

Early prototypes of the turbine used a 100mm diameter wooden hub with the blades bolted into slots around the perimeter. This design was immensely strong and durable but difficult to produce without complex jigs and precision woodwork; hence the redesign of the turbine using pressed aluminium throughout and pop rivets to assemble the blades to the hubs.

hub

The hub comprises two 100mm diameter circular plates of 2mm thick aluminium as shown in Figure 7 [turbine hub plate].

The hub plates can be made from 1mm galvanized steel.

I have had no problems with mixing aluminium and galvanized steel, but I know that it is better not to mix these materials. If convenient, make hubs and blades from the same material.

Central hole: 8mm in front hub plate; 12.5mm in rear hub plate

If you leave the square section on the turbine spindle, then the front hub plate needs a square hole, (8.5mm side). See section on turbine construction

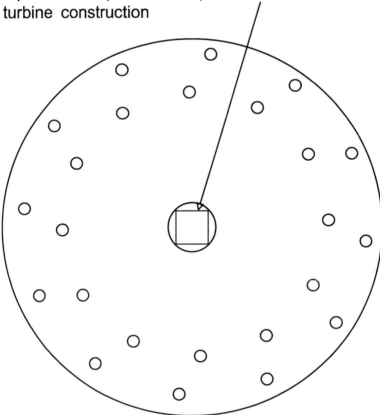

Figure 7: turbine hub plate

12 sets of holes set symmetrically about
the hub plate

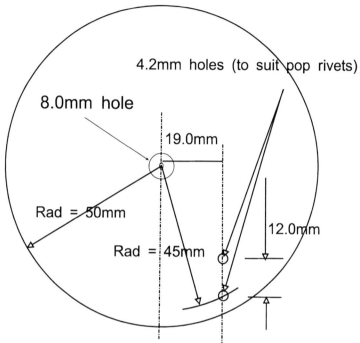

4.2mm holes (to suit pop rivets)

8.0mm hole

19.0mm

Rad = 50mm

12.0mm

Rad = 45mm

Figure 8: hole positions in the hub plates

The reason for not mixing metallic materials is well documented elsewhere but briefly it is as follows:

Where two dissimilar metals are in close proximity in a moist environment, any acidic fluid between the metals will create an electrolytic or battery effect. The acid is usually dissolved carbon dioxide in rain but in coastal environments it can also be dissolved sea salt. The result is corrosion of one metal in preference to the other, at an accelerated rate.

Figure 8 [hole positions in the hub plates] shows how to position the holes to be drilled in the hub plates, and Figure 9 [turbine hub place spacing] shows how the hub plates are oriented with the holes ready to line up with the blades.

Provided that the plates are identical and that the plates are parallel, the exact distance between the plates is unimportant

However, the length of the turbine spindle is affected by the distance between the plates

Distance between plates is approx 34mm

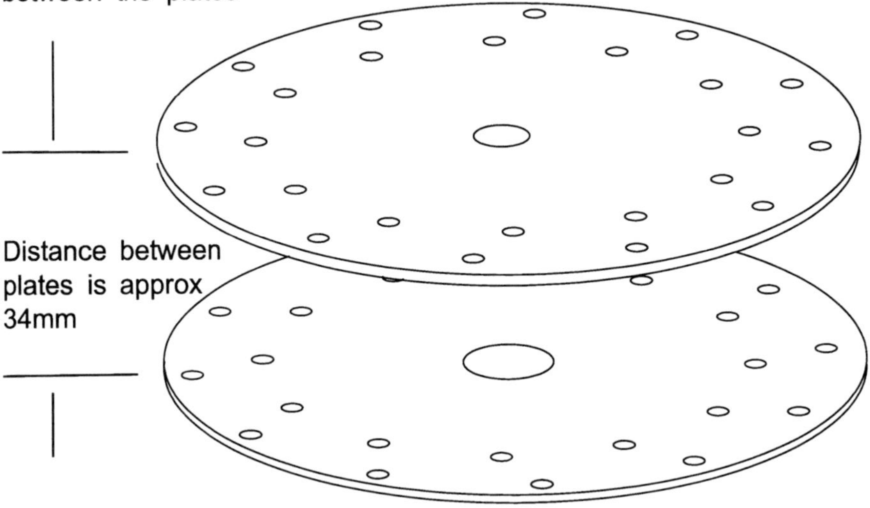

Figure 9: turbine hub plate spacing

Note that in the earlier design, the rear hub plate has a larger central hole to accept the front of the spindle bush. The hub plates are kept apart by the blades with the assistance of a central wooden block, the turbine hub spacer. The turbine hub spacer is made from a section of broom handle and may be used to serve the added purpose of locking the turbine to the drive spindle. (See Figure 12: turbine hub spacer).

This design has not been changed since the first rotor was made in 1994. it is very successful and easy to build.

blades

The blades are cut from 2mm thick aluminium (or 0.7mm thick galvanised steel sheet if you can work it).

I prefer to use aluminium because it is easier to work by hand, but the galvanised steel sheet is fine, especially if that is what is available.

The pattern for blades is shown in Figure 10 [turbine blades], and Figure 11 [hole positions in the turbine blades].

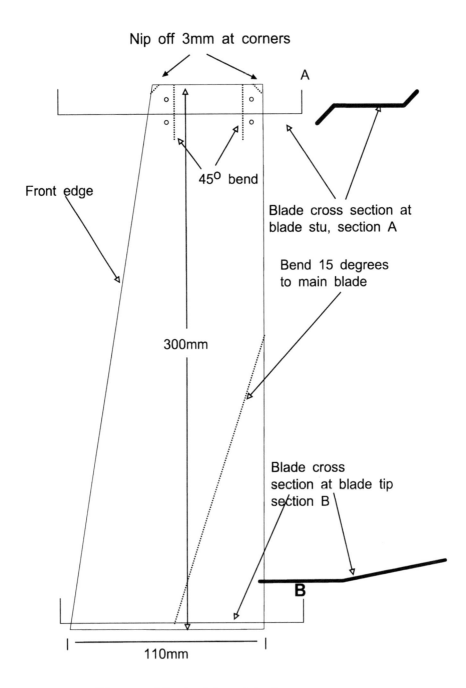

Nip off 3mm at corners

A

Front edge

45° bend

Blade cross section at blade stu, section A

Bend 15 degrees to main blade

300mm

Blade cross section at blade tip section B

B

110mm

Figure 10: turbine blades

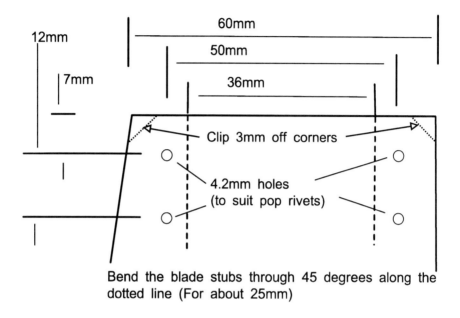

Bend the blade stubs through 45 degrees along the dotted line (For about 25mm)

Figure 11: hole positions in turbine blades

The forming process is simple and only needs a vice with 10cm jaws. It is essential that the holes in the blades line up with the holes in the hubs; this requires that the marking, cutting, drilling and bending is done with great accuracy.

The most important point is that all the blades are identical and that the hole spacing of the blades is the same as that of the hub plates. If the width between the pairs of holes at the stub of the blade is a millimetre wider or narrower than drawn, i.e. 50mm, then the turbine hub plates will be further apart or nearer together (but this is not a problem).

wood spacer

This is a piece of 25mm diameter dowel axially drilled 8mm and cut square to length 29mm. It is important to make sure that the ends of the spacer are perpendicular to the axis.

See Figure 12 [turbine hub spacer].

I have not built a rotor without the hub spacer but I reckon it may be OK, as the assembly is very strong indeed.

If you decide to build a rotor without the spacer it is certainly lighter but if it is NOT strong enough or disfigured, it is difficult to retro-fit the hub spacer.

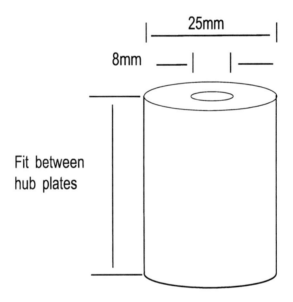

Figure 12: turbine hub spacer

assembly of turbine

Assembly of the turbine is simple. The blades are attached to the hubs using 4.2mm pop rivets. (Early versions used 3.2mm pop rivets but while the structure was generally OK, I think the extra strength of the 4.2mm rivets is worth it.)

The blades are not symmetrical, and care must be taken that they are all installed with the angled leading edge facing forward. In the old design, the hub plate with the larger central hole (12.5mm) faces to the rear. The new design has the same sized holes front and rear.

Start by loosely bolting the two hub plates together with the wooden hub spacer between them; make sure that the holes in the hub plates do not line up, (if they do, just turn one hub plate upside down).

The order of fitting the blades to the hubs is optional, but it may be simpler to fit two opposite blades to one hub (180° apart), then rivet each blade to the other hub. Doing it this way will give a sturdy structure to add to.

Do not be tempted to use rivets in oversize holes or widen holes to accommodate badly-prepared hubs or blades: if the rotor fails because of one blade and subsequently loses that blade, the whole turbine will be unstable and may fly to bits. If it does this, the remainder of the turbine will be seriously out of balance and you may subject the rest of the structure to some interesting loads, even if only briefly!

spindle

The design of the spindle and turbine head has been improved since the first book. The improvements have resulted in a system that allows the rotor to be removed from the spindle without having to remove the spindle from the turbine head; and it also turns out to be cheaper.

Figure13 [turbine spindle] shows the old spindle design made from an M8 coach bolt. This design does work and if you have these bolts to hand, consider using them.

You still need to fix a sleeve to fit the bearing in the turbine head - see Figure 14 [spindle with sleeve].

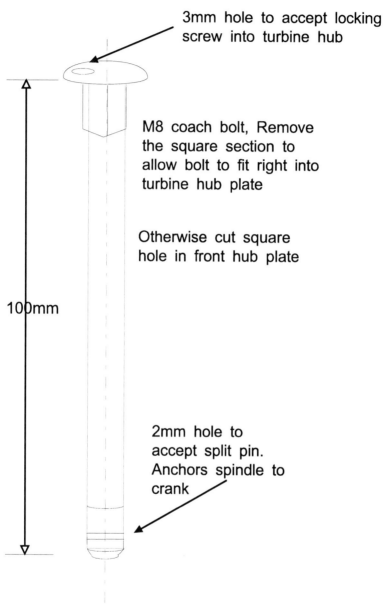

3mm hole to accept locking screw into turbine hub

M8 coach bolt, Remove the square section to allow bolt to fit right into turbine hub plate

Otherwise cut square hole in front hub plate

100mm

2mm hole to accept split pin. Anchors spindle to crank

Figure 13: turbine spindle

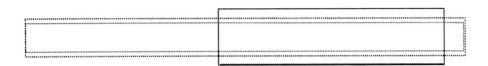

Figure 14: spindle with sleeve

I now make the spindle from a 100mm length of M8 threaded rod. The amount .of machining is reduced and the only slight complication is that the task of brazing the threaded rod into the sleeve is harder than brazing a straight-sided bolt into the sleeve. The dimensions of the sleeve are unchanged (8mm ID, 12mm OD * 51.0mm long) but the sleeve is now made of steel tube.

If you like, you may turn the spindle from one length of M12 silver steel by simply turning down the ends and threading each end M8. This however, is not a cheap option.

If you make a geared pump, the spindle needs to be some 5-7mm longer at the rear end as, unlike the crank which is brazed to the spindle, the sprocket is bolted to the spindle.

spindle sleeve

The spindle sleeve is steel. I hope that it is apparent that if convenient sizes of stock tube are to hand, this component, and several others, can be made by simply sawing off suitable lengths of tube and brazing them together, otherwise you must turn the sleeve from rod and bore to M8 ID.

The spindle sleeve is locked to the spindle; the spindle is not supposed to rotate in the spindle sleeve as might be thought.

crank

If you are going to build a geared pump as described later, you will NOT need a crank. The connecting rod will be attached to a pin directly fixed to the larger sprocket.

You do need a crank if you are building the direct drive machine.

The crank is made from steel and is shown as part of Figure 15 [spindle and crank assembly]

It is not necessary to make the crank symmetrical about the spindle axis, but the extension opposite the connecting rod end can be used to carry experimental counterweights.

These small turbines tend to run a lot faster than their larger counterparts and for this reason balance is more important. Even if you do not fit counterweights to the far end of the crank, the fact that it is extended past the spindle axis does help to balance it.

Since the new crank is steel, it can be less than the 6.5mm shown in the figure. I am using 20.0mm * 5.0mm black bar steel and it is fine.

Also since the crank is now brazed to the spindle there is no longer a requirement for the split pin shown in the figure in the first design.

Figure 15 also shows the complete assembly of the spindle, sleeve and crank. At the rear end, the crank is screwed onto the spindle and the spindle is brazed to the crank. This means that the spindle, crank and sleeve are one permanent assembly and to remove the crank from the turbine head implies removal of the turbine blades; but the improved strength this new design gives is worth it.

You may disagree and decide to use a nut plus lock-nut to secure the crank to the spindle, enabling you to remove the crank without removing the rotor. Horses for courses!

You will, however, need to extend the length of the spindle to accommodate the extra lock-nut.

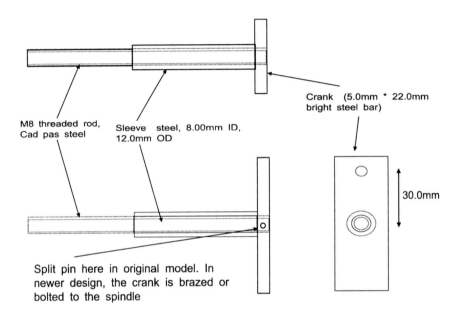

Crank (5.0mm * 22.0mm bright steel bar)

M8 threaded rod, Cad pas steel

Sleeve steel, 8.00mm ID, 12.0mm OD

30.0mm

Split pin here in original model. In newer design, the crank is brazed or bolted to the spindle

The crank has two holes - one tapped M5, the other M8. Both these holes must be exactly perpendicular to the flat surface of the crank.

Having brazed the crank to the spindle and the spindle to the sleeve, the back surface of the crank may be trued in a lathe. But if it requires more than 0.25mm material to be removed, it may be better to start again!

NB: the crank must not wobble as the spindle turns; the axis of the con rod pin must be parallel to the spindle axis.

Figure 15: spindle and crank assembly

(In the old design you could remove the crank from the spindle, leaving the turbine on the spindle.)

The turbine rotor is slid onto the spindle and must be locked there securely.

The important point is that the turbine rotor must be very firmly attached to the spindle. If it starts to turn on the spindle, it will rapidly expand the bore of the rotor and become unbalanced. This happened to one of my turbines and it dashed itself to bits against the tower. This process happened very noisily at 0400 hours and I had to climb the tower in a strong wind wearing dressing gown and slippers to dismantle it! Fortunately the neighbours thought the noise was caused by lots of dogs tipping over lots of dustbins (which is what it sounded like).

If you decide to go for the newer, geared pump, you replace the crank with a 16-tooth cog (sprocket) as shown in Figure 16 [spindle and small chain cog assembly].

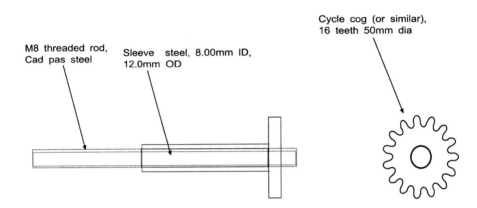

The hole in the cog is hopefully suitable for the M8 thread on the rear of the spindle. If the spindle is too small then you need to make a small sleeve to take up the difference. If the M8 thread is too large (unlikely), then bore out the cog. Try to avoid reducing the M8 thread as the load is too much for a small thread.

Figure 16: spindle and small chain cog assembly

connecting rod

The connecting rod (con rod) is steel or duralumin (see glossary); it can be fashioned from a length of 3mm thick sheet or bar. I use 13.0mm * 5.0mm black bar steel. The exact dimensions are unimportant but the length must obviously be compatible with the length of the push rod. Reducing the length of the push rod and the connecting rod by too much will reduce the efficiency of the turbine assembly. Play around with the recommended sizes by all means, but you need a good reason to shorten them by much.

See Figure 17 [connecting rod].

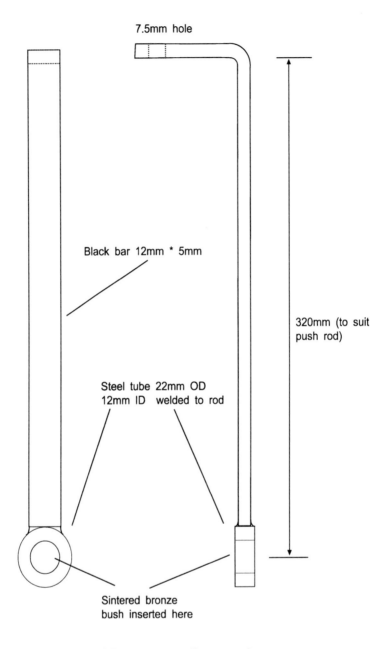

7.5mm hole

Black bar 12mm * 5mm

Steel tube 22mm OD
12mm ID welded to rod

320mm (to suit
push rod)

Sintered bronze
bush inserted here

Figure 17: connecting rod

There is a bearing hole at the lower end of the con rod and if the con rod is aluminium, this hole must be reinforced. To do this, I used a suitable sized nutsert (see resources) with the thread bored out. The top end of the con rod is bent over to sit over the top of the push rod.

When using a steel con rod, I have started to fit 'oilite' bushes (see glossary) into the lower end of the push rod. I think it is an improvement but since this is a cheap component to make, you may well decide to leave it unbushed and simply replace it more frequently.

(I recommend the steel con rod with an oilite bearing.)

push rod

The push rod is a length of steel rod 8.0mm diameter. See Figure 18 [push rod]. The rod must be turned down to take a male M5 thread for the long transmission rod. In the new design the top end is now turned down to take an M5 thread for 30mm. The detail in the top right of the drawing shows how I introduced a shock absorber into the drive from con rod to push rod; the springs will absorb excess movement should the turbine be rotating too fast. I cannot specify the springs and they are not essential.

Detail

If the rotor moves too fast in high winds, it will try to raise water too quickly and the transmission system will fail.

In normal use, the springs do not compress, and all the energy of the rotor is transmitted to the pump. IF, however, the wind is high, the springs will compress and absorb most of the rotor energy; the pump will not move water but at least the system is saved.

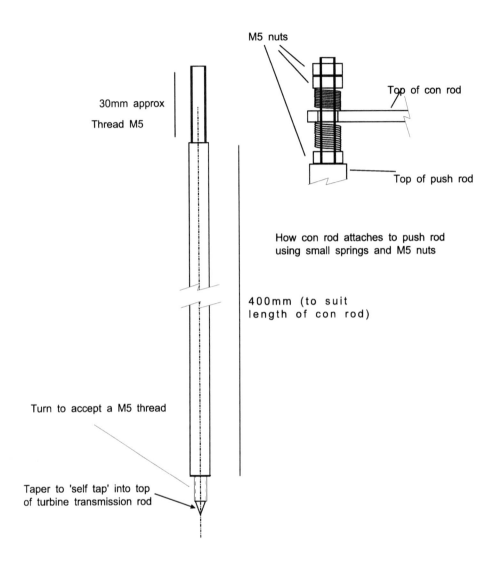

M5 nuts

30mm approx

Thread M5

Top of con rod

Top of push rod

How con rod attaches to push rod
using small springs and M5 nuts

400mm (to suit
length of con rod)

Turn to accept a M5 thread

Taper to 'self tap' into top
of turbine transmission rod

Figure 18: push rod

turbine head assembly

This is the only part of the wind pump where a lathe really makes life easy. Without a lathe, the components of the turbine head assembly can be fabricated from suitable brass or steel tube stock and attached using brackets (instead of machining to close tolerances and hard soldering).

This is one of the components that has been improved most since my original design.

head (original design)

I have included this description (and the associated drawings) for completeness. The new design is substantially different from the original and uses different materials. However, it may suit some people better than the latest design. Study this design, and the latest one, and decide for yourself which one suits your own engineering desires and capabilities.

ALL the designs of the head assembly are interchangeable.

See Figure 19 [turbine head components].

You can see how the components fit together in Figure 4.

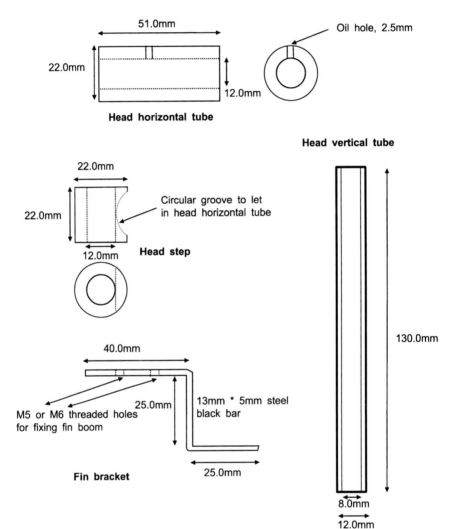

51.0mm

22.0mm

12.0mm

Oil hole, 2.5mm

Head horizontal tube

Head vertical tube

22.0mm

22.0mm

12.0mm

Circular groove to let
in head horizontal tube

Head step

130.0mm

40.0mm

25.0mm

M5 or M6 threaded holes
for fixing fin boom

13mm * 5mm steel
black bar

25.0mm

8.0mm

12.0mm

Fin bracket

Figure 19: turbine head components, original design

This is constructed of two steel tubes welded together at right angles. A circular groove is cut in the vertical short tube (the head vertical tube) to let in the horizontal tube (the head horizontal tube).

The spindle sleeve runs inside the head horizontal tube, so the inside bore of the head horizontal tube needs to match the outside diameter of the spindle sleeve; the actual size is immaterial so long as the wall thickness is greater than 3mm. A 2.5mm hole drilled down into the bore of the head horizontal tube allows oil to be fed to the spindle bush.

The head vertical tube has two functions: the inside is the bearing surface for the push rod which slides vertically inside; and the base of the tube sits in the support tube that is attached to the wind pump tower. The bottom surface of the head step rests on the top surface of the support tube. This allows the whole turbine assembly to rotate on a vertical axis.

See Figure 20 [turbine head assembly].

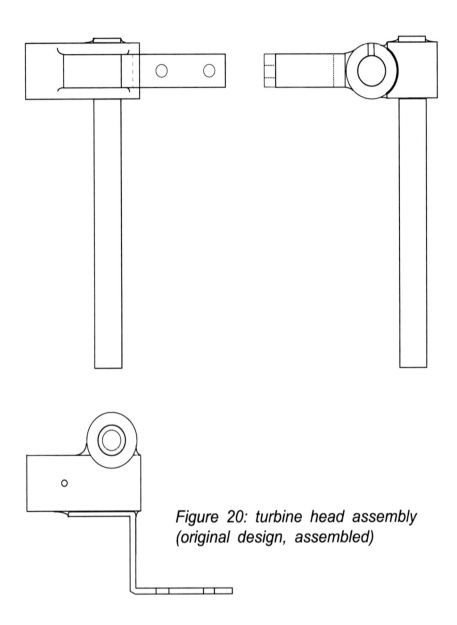

Figure 20: turbine head assembly
(original design, assembled)

The tail fin bracket is welded to the left-hand side of the horizontal tube (viewed from the back). This bracket is simply a right-angle 'Z' shape in the original design, but if you are going to have a go at a self-furling turbine (described later), then this bracket needs to be more complex.

head (intermediate and latest designs)

The original design of all the components for the head focused on lightweight construction; I am still very conscious of the need to keep the weight down, but since there has never been a failure of the tower structure that could be attributed to that sort of load, I have experimented with stronger and heavier head designs.

Also, with the advent of a geared option for the drive, AND self-furling, more strength was needed for the head - hence this exploration.

Instead of welding two tubes together at right angles, I decided to fabricate the head from one piece of metal.

The result is a very strong head made from a solid oblong of steel, bored horizontally along its longest dimension to take the spindle, and bored vertically to take the head vertical tube.

The resultant configuration of spindles and push rods etc. has not changed - it just looks a bit different.

See Figure 21 [turbine head components (square head)].

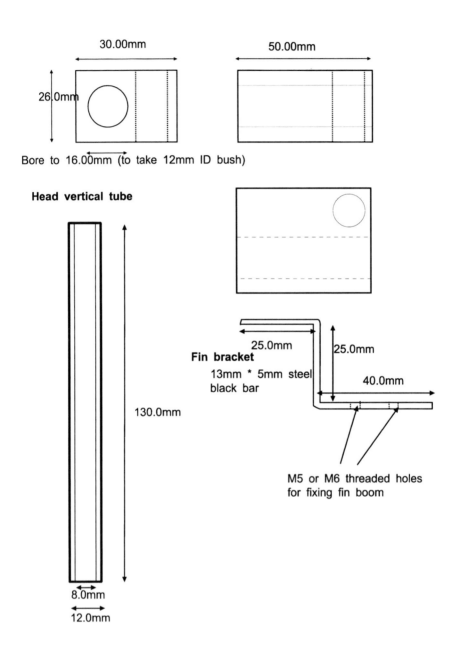

30.00mm

50.00mm

26.0mm

Bore to 16.00mm (to take 12mm ID bush)

Head vertical tube

130.0mm

8.0mm

12.0mm

25.0mm

Fin bracket

13mm * 5mm steel
black bar

25.0mm

40.0mm

M5 or M6 threaded holes
for fixing fin boom

Figure 21: turbine head components (square head)

The only fabrication problem that this introduces is that some people may be able to make the old tube-based head from existing tube stock, but are unable to bore a 12-15mm hole through a solid piece of steel 50mm long.

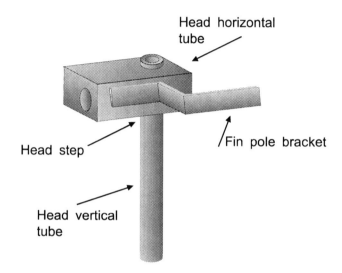

Figure 22: assembled square turbine head

turbine tail stick and tail fin

The stick can be made of any light material provided it is not too flexible: a 12mm - 20mm plastic water pipe is OK but if it is available, 12mm OD aluminium tube is best.

See Figure 23 [tail fin bracket and attachments] for details.

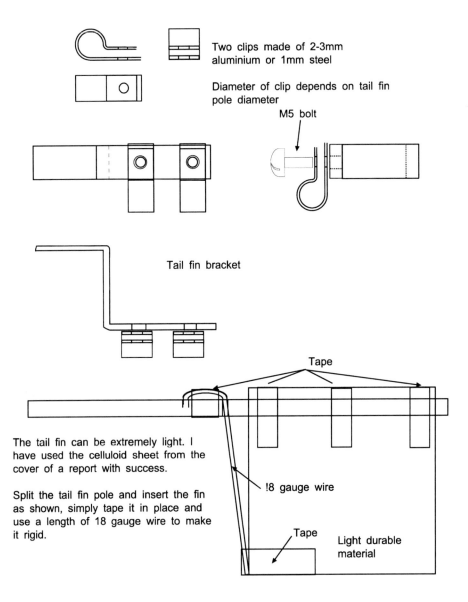

Two clips made of 2-3mm aluminium or 1mm steel

Diameter of clip depends on tail fin pole diameter

M5 bolt

Tail fin bracket

Tape

The tail fin can be extremely light. I have used the celluloid sheet from the cover of a report with success.

Split the tail fin pole and insert the fin as shown, simply tape it in place and use a length of 18 gauge wire to make it rigid.

!8 gauge wire

Tape

Light durable material

Figure 23: tail fin brackets and attachments

This is another area that has been improved in strength and ease of construction since the first book.

The stick is clamped to the tail fin bracket, which is welded to the side of the turbine head assembly. The clamps can be made from strip aluminium 3mm * 15mm bent into a letter P which is squeezed around the stick by bolting through the leg of the P to the bracket.

The fin needs to be very light. The shape is not critical but for maximum rigidity, it needs to be fairly square. An oblong fin 200mm * 250mm attached to a 400mm stick works fine.

mast

The mast is probably where the greatest compromise is tolerated between requirement, design and implementation. You need to get your turbine into a good clean air flow; for this you probably need a tower 20 metres high! OK, let's settle for 5-6 metres - this compromise is for the neighbours as much as for your bank balance.

A 7-metre scaffold pole buried 2 metres into the ground (concrete) is OK, except that it may not be aesthetically pleasing to you - and how do you get up to the top to do repairs or modifications? One plan is a small trellis tower of timber or light steel tube to about 3 metres, with a flat 500mm square top to stand on. The mast is then fixed vertically as part of the trellis and extending up a further 2-3 metres.

A suitable pole attached to the side of a single-storey flat-topped building can work, provided that the building does not have too much effect on the clean air flow.

See Figure 24 [tower using scaffold pole and guy ropes], and Figure 25 [trellis tower], which may prompt further thoughts on solving this problem.

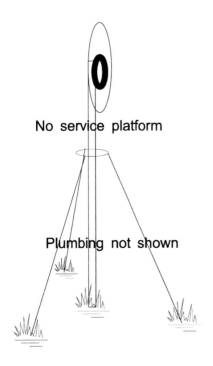

No service platform

Plumbing not shown

Figure 24: tower using scaffold and guy ropes

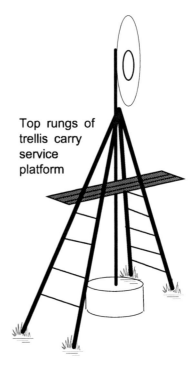

Top rungs of trellis carry service platform

Figure 25: trellis tower

The best build I have come up with is shown in Figure 26 [tilting turbine tower].

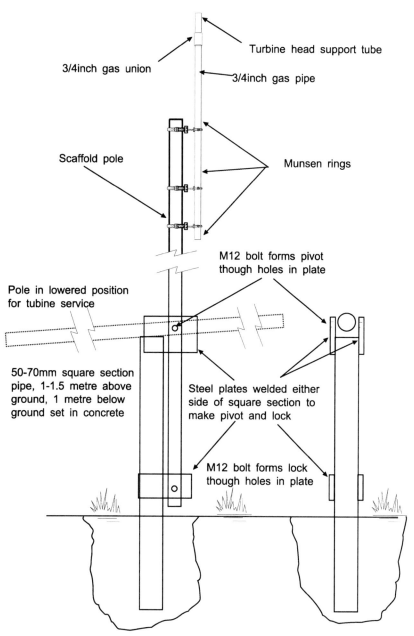

3/4inch gas union

Turbine head support tube

3/4inch gas pipe

Scaffold pole

Munsen rings

M12 bolt forms pivot though holes in plate

Pole in lowered position for tubine service

50-70mm square section pipe, 1-1.5 metre above ground, 1 metre below ground set in concrete

Steel plates welded either side of square section to make pivot and lock

M12 bolt forms lock though holes in plate

Figure 26: tilting turbine tower

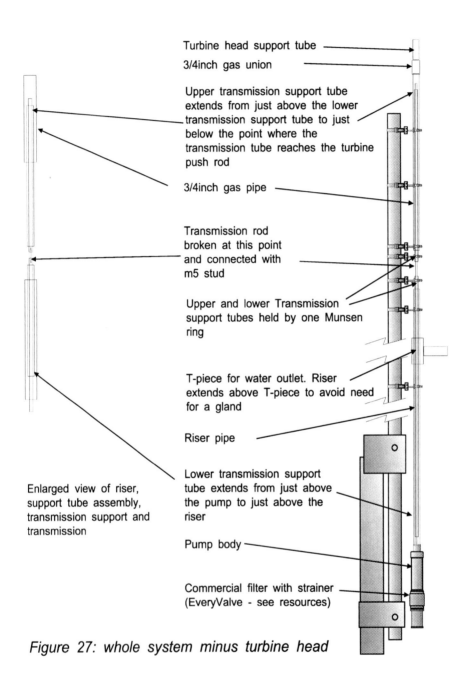

Turbine head support tube ──────

3/4inch gas union ──────────

Upper transmission support tube
extends from just above the lower
transmission support tube to just
below the point where the
transmission tube reaches the turbine
push rod

3/4inch gas pipe ──────

Transmission rod
broken at this point
and connected with
m5 stud

Upper and lower Transmission
support tubes held by one Munsen
ring

T-piece for water outlet. Riser
extends above T-piece to avoid need
for a gland

Riser pipe ──────

Enlarged view of riser,
support tube assembly,
transmission support and
transmission

Lower transmission support
tube extends from just above
the pump to just above the
riser

Pump body ──────

Commercial filter with strainer ──────
(EveryValve - see resources)

Figure 27: whole system minus turbine head

I include Figure 27 at this point to give an overview of the entire structure below the turbine itself. There is no right way to design or create this section of the build, but if you are following my design closely, this schematic will give an idea of the whole thing.

support tube

Figure 28 [support tube assembly] shows the new support tube. I make this from a length of the 22mm OD, 12mm ID tube (the same as I used to use for the original horizontal head tube). This is attached to the tower at the top and allows the head vertical tube to rotate freely about a vertical axis. The support tube may be attached to the mast in many ways, but the important point is that the whole fixture must not be too wide or the turbine blades will hit it.

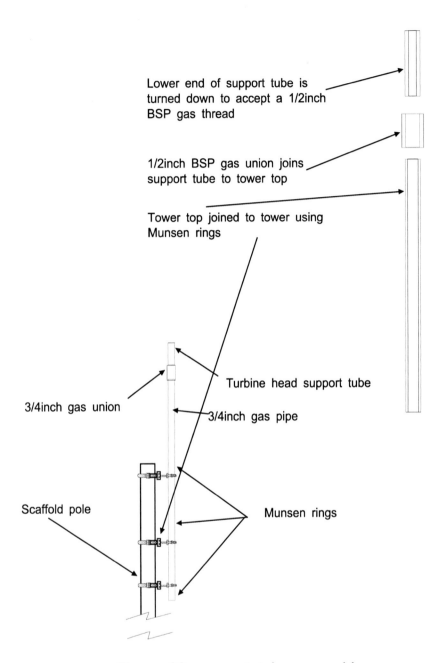

Lower end of support tube is turned down to accept a 1/2inch BSP gas thread

1/2inch BSP gas union joins support tube to tower top

Tower top joined to tower using Munsen rings

Turbine head support tube

3/4inch gas union

3/4inch gas pipe

Scaffold pole

Munsen rings

Figure 28: support tube assembly

Bear in mind that the trailing edge of each blade passes within 30mm of the centre line of the support tube, so the maximum diameter of the top 400mm of the mast assembly must be less than 60mm.

As you can see, the support tube is threaded so that it screws straight into a 3/4inch steel gas pipe union which in turn screws onto a 1 metre length of gas pipe. The pipe is firmly attached to the top 300mm of the scaffold pole using Munsen rings.

I don't know how international Munsen rings are, but each ring consists of two halves that can be tightened around a pipe, each complete ring is attached to another ring by a short length of M10 studding. In this way, two pipes of similar or different diameters may be firmly attached to each other; the pipes can be parallel and the distance between them controlled by the length of the studding.

I apologise for the mixing of dimensions in this section but all the gas pipe sizes are imperial with BSP threads on the pipes.

transmission

The transmission covers the mechanism between the bottom of the push rod in the turbine head and the top of the pump. Now in experimental wind pumps frequent changes may take place in the exact distance between the pump and the turbine so a cheap method has evolved which allows quick modifications to be carried out.

Most of the long transmission rod is a 6mm OD * 2mm wall PVC tube.

See Figure 29 [transmission rod]

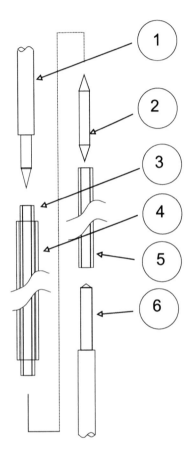

1 Bottom end of push rod
2 M5 stud
3 Turbine transmission rod
4 " " " sleeve
5 Pump transmission rod
6 Pump stem

The pump transmission rod should be long enough to reach just above the pump stand pipe. This allows you to disconnect the pump from the turbine at ground level, while leaving the pump and turbine where they are.

The transmission sleeve is clamped to the tower structure and extends from a few inches below the turbine head to a few inches above the pump stand pipe.

Figure 29: transmission rod

This is very light but far too flexible, so it runs in a similar PVC tube with a 9mm bore. The outer tube is rigidly held in position between the turbine and the pump by clamps.

I now use Munsen rings to hold ALL pipe work whether it is the transmission rod guide or the main riser.

The transmission rod sleeve must be firmly attached to the mast and aligned both with the pump at the base of the tower and with the push rod at the top.

The push rod in the turbine head is threaded M5 at its lower end, and this screws straight into the top of the transmission rod.

It is quite likely that after some experimenting, you will find that the transmission rod has been shortened and lengthened several times; as a result there may be several joins in the transmission rod. Each join requires a 20mm length of M5 studding to be screwed into the transmission rod.

The bore of the 6mm tube should be tapped to a depth of 10mm to allow the studding to penetrate well.

If the studding is filed to a 60° point, it can be used to tap the bore of the transmission rod like a self tapping screw. However, I would only do this in an emergency.

The top of the piston rod is joined to the transmission rod in a similar manner - see the pump details in Figure 35.

pump body

This is a lift pump, the body of which is constructed entirely from off-the-shelf plumbing fixtures, available from most large DIY retailers. Once you see the general design, you will probably be able to adapt it to suit the fittings that are available to you locally.

In the original, the manufacturer of the bits I used is 'OSMA' (see resources), and I have shown their codes for the fittings used in the design in Figure 30 [old pump body (plastic)].

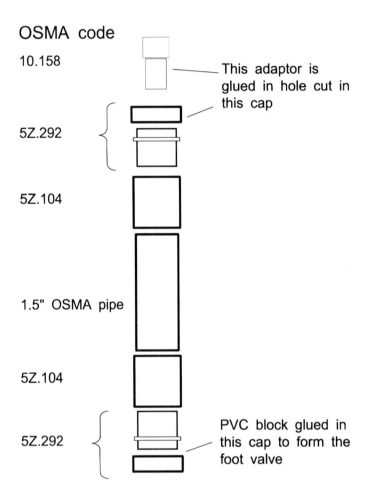

Figure 30: old pump body (plastic)

Figure 31 [home-made foot valve] shows the detail of the OSMA foot valve. I have long since changed this part of the system and in normal use I recommend that you do NOT use a plastic-walled pump. If however, you are happier working with the OSMA components, and the pump is for very light loads, it may be worth experimenting.

OSMA part 5Z.292 is the basis for the plastic foot valve

M6 bolt with nut + lock-nut to loosely secure PVC disc over holes in OSMA cap

Solid PVC disc, able to rise and fall on the M6 bolt

Reinforced plate glued to base of the OSMA cap

OSMA cap and reinforced plate has 6 holes to allow water ingress during pump upstroke

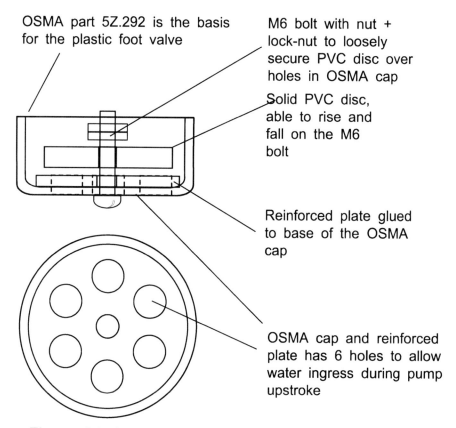

Figure 31: home-made foot valve

Another disadvantage of the plastic bore pump (which I only discovered after an embarrassingly long time), is that the bore is not uniform! So a piston that is snug in one part, will leak in another.

I do not include the detailed description of the original plastic-walled pump, although the figures should be self-explanatory.

You can still use the top and bottom OSMA caps, as shown in Figure 30, to make the next pump (steel) but even here the design has evolved, and if you use a steel body for the pump and a steel riser, you may as well capitalise on the improved strength of a steel top cap and metal foot valve. I simply mention in passing that ALL the bits are still compatible for plastic or steel pump until we move onto the larger-bore pump for the geared turbine.

The pump body is made from a 100mm length of 1.25inch steel pipe. (I apologise again for mixing dimensions but this is a standard threaded gas pipe size here in the UK.)

If possible, you should bore out this tube by a small amount just to remove the weld seam from the inside of the pipe. If this cannot be done, it will mean that the piston is NOT a snug fit in the bore. The pump will still work, though less well. As a compromise, you might try to file off all the 'high' spots from the weld.

If you can use a lathe to remove the weld, there are a few tips to make it easier, especially if you don't have a 'steady': off-set the pipe in the chuck, and make the weld seam closer to the axis of the lathe. The weld high spot will get the attention of the cutting tool well before the rest of the circumference of the pipe. If the lathe has a four-jaw chuck this is all in the setting up, but if you have only got a three-jaw chuck, you need a 1-2mm spacer between one jaw and the outside of the pipe next to the weld seam.

I buy the pipe already threaded and to protect the thread I screw it into a union and grip the union in the lathe. This always seems to result in an unavoidable offset and it is simply a matter of making sure the offset is in the right place!

If you have a lathe, you should understand this already, and if not, ignore my meanderings and file the weld out - it still works.

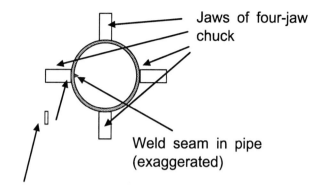

Jaws of four-jaw chuck

Weld seam in pipe (exaggerated)

1-2mm spacer to offset pipe from lathe centre if you only have a three-jaw chuck

Figure 32: offsetting pipe in lathe

The diameters of the various bits inside the pump will be governed by the size of fixtures available, but the pump body should be long enough to allow the piston to travel 70mm without striking the ends of the body. This 70mm may be different if you use a different size crank, or change the position of the point of attachment of the con rod on the large sprocket of the geared pump.

If you are likely to adjust the turbine height or pump depth much during experiments, you would be wise to allow 50mm above and below the stroke of the piston. Failure to do this may result in the piston hitting the foot valve or the top of the pump body. 50mm may seem a large tolerance but you may be adjusting the vertical position of the pump when you cannot even see it under the water.

Allowing for the 60mm piston stroke, this gives a pump body length of 160mm! If you use the bored-out steel pipe for the pump body, you must take care. My own lathe is not really suitable for boring lengths greater that 100mm unless using a reamer or a steady, so I arrange for the pump body to be JUST long enough - i.e. 80-90mm. Be careful, as this boring process is potentially dangerous. If you are unsure of your skill, get the bore done professionally. It will not add greatly to the cost - a lot less than a lost eye, limb or even life if the pump body leaves the lathe at speed.

Experience has shown that the parts of the pump that the piston is working against come under quite a substantial load, whilst the actual riser pipe is not very loaded at all. For this reason I have made all the fixed (clamped) parts of the pump from steel in the new system.

In particular, the riser, from the top of the pump up to ground level, cannot be supported as the supports need to be above ground level, so it makes sense to use a steel riser to this point at least.

We need to attach a steel pipe from the pump to the riser (OSMA overflow pipe). It turns out that the union thread of the OSMA overflow pipe is the same thread as a 3/4inch BSP thread.

So all we need to do is reduce the pump bore using a 1.25inch - 0.75inch female-female reducer and then using a length of 0.75inch steel pipe for the clamped part of the riser, connected to the OSMA riser using a 20mm union.

See Figure 33 [new design for pump body using steel reducer for increased strength].

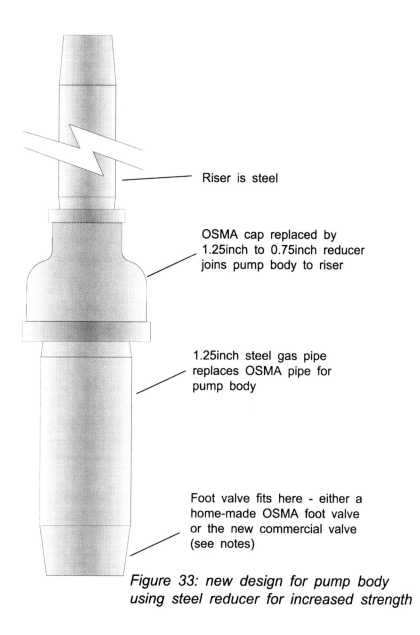

Riser is steel

OSMA cap replaced by
1.25inch to 0.75inch reducer
joins pump body to riser

1.25inch steel gas pipe
replaces OSMA pipe for
pump body

Foot valve fits here - either a
home-made OSMA foot valve
or the new commercial valve
(see notes)

*Figure 33: new design for pump body
using steel reducer for increased strength*

If you wish, you can still use the OSMA parts as shown in
Figure 31 to join the riser to the pump body, but I think the
new method just described is better.

I will move onto the foot valve, starting with the old design. Even if you decide not to make your own foot valve, I think this next bit may be interesting if you need to know the how and what of foot valves.

DIY foot valve

The design in Figure 31 does work, but needs regular servicing. The commercial foot valve I have sourced here in the UK is fantastic. So much so, that I have stopped building my own. There are undoubtedly users who by desire or necessity will build their own, so a bit of background on foot valve construction and how they work may help.

The cap at the bottom of the pump (foot valve) is a bit trickier than the top cap of the pump: it has to be reinforced by gluing a solid 5mm-thick PVC round block to the inside of the cap.

The block has holes bored in it through which the pumped water flows; the block is also the main sealing surface of the foot valve and detail of this component is covered in the section on valves. (A similar block performs a similar task as part of the piston assembly.)

When attaching the block, precision takes second place to strength, but remember that one of the most critical factors in maximising the efficiency of the pump is the seal in the foot valve, so try to ensure that the flap-valve sits flat and square in the bottom of the pump body. Also remember that the threaded pipe body of the pump MAY extend into the foot valve far enough to interfere with the valve block, so the diameter of the block must be less than the ID of the pump body.

In early models, I made the flap valve for the foot valve from metal, usually 3mm thick brass. Sometimes they worked for ages but sometimes they would start to leak and needed to be changed. Later I used a disc of rubber with a disc of metal above it to spread the load; the rubber can be from a car inner tube or a small bit of medium-duty butyl pond liner. I have even tried washing-up gloves (but don't bother - they are not strong enough).

The operation of both valves relies on the fact that a metal, plastic or rubber disc (the valve plate) sits over the holes in a larger PVC valve seat, preventing water from flowing through.

When the water pressure changes on one side of the valve, the plate is lifted allowing the water to flow. The plate is restricted in its movement by a loose-fitting bolt or pin passing through to the base plate. This simple process is more obvious when looking at the sketches, but to watch the valves in action is best of all. You can buy a child's bathroom toy lift pump that operates on this principle; it has a clear perspex tube so you can observe the mechanism.

The greater the head of water above the pump, the better the valves work.

There is an optimum amount of movement for the valve plates. Let's look at this movement in the foot valve.

The amount of water flowing through the open valve is obviously related to the size of the holes in the valve seat, so lifting the valve plate by more than a few mm serves no purpose. If the valve plate only lifts a short distance (less than 1mm say), then the full flow potential of the valve is not achieved. If however, the valve plate is allowed to rise by 10mm say, then although the valve would be fully open during the lift cycle, it will remain open for a while after the lift cycle while the valve plate falls back to the valve seat. During this time, water will flow back through the valve.

If the valves do not open enough, the pump feels as if it is stalling; if the valves open too much, the pump becomes very noisy and inefficient.

If you calculate the cross-sectional area of the holes in the valve and allow twice that area in the vertical gap between the valve seat and the valve plate, it will work fine.

This may require further explanation. When the valve plate lifts to admit water through the holes in the bottom of the valve, this water then needs to flow between the base of the valve and the valve plate and then between the side of the valve plate and the side of the pump.

If the valve plate only lifts say 0.01mm, then obviously this wouldn't allow as much flow as if it lifted 10mm. The idea is to make the allowance of flow similar to that provided by the holes in the base of the valve.

Similarly, the gap between the edges of the valve plate and the sides of the pump wall should also allow the same flow.

You also need to calculate the diameter of the valve plate. The area of annulus between the outside of the valve plate and the inside of the pump wall needs to be at least as large as the area of the holes in the valve seat. In any event the valve plate should just cover the holes in the valve seat.

The holes in the foot valve and the piston are shown as a ring of 6 holes in Figure 31.

The piston has a valve which operates in an identical manner; the construction is similar but the seal is not quite so critical since the piston is going to leak anyway.

why use a home-made foot valve?

A last comment before we leave home-made foot valves. I have tried to ensure that all the components of this system can be made from scrap or recycled bits and pieces, and for the most part, I think I have managed to do this. My reason is not simply economy but also to prevent the user from becoming reliant on some manufacturer. Many of the enquiries for this system have been from areas where NO engineering supplies would be available; people in these locations can still build their own valves.

These home-made valves are very effective for the piston, but I have found that a commercial foot valve is worth the expense. I have found a UK supplier who sells a brass foot valve with a nylon valve seat and a stainless steel strainer (filter); this valve has a 1.25inch BSP female thread and is VERY robust. It is also far superior to anything I have made. (Cost about £15-£20).

The UK supplier is listed in 'resources'.

Figure 34 [pump assembly with commercial foot valve and strainer] shows the commercial foot valve fitted to a steel pump body and an OSMA top cap.

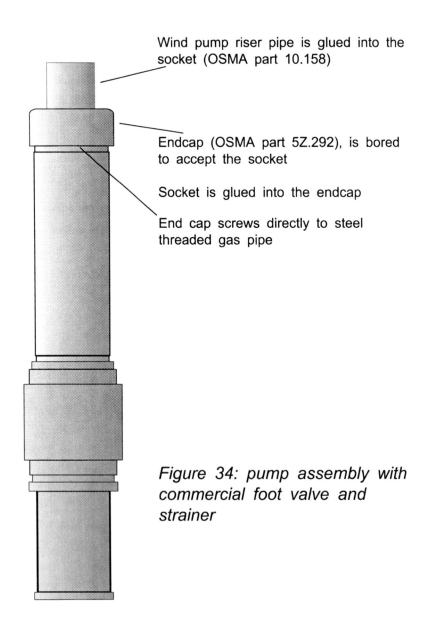

Wind pump riser pipe is glued into the socket (OSMA part 10.158)

Endcap (OSMA part 5Z.292), is bored to accept the socket

Socket is glued into the endcap

End cap screws directly to steel threaded gas pipe

Figure 34: pump assembly with commercial foot valve and strainer

piston and stem

The piston stem is a piece of 7-8mm steel rod about 100 to 200mm long (actual size does not matter but a long stem will increase the weight).

The piston is a disc of PVC sheet about 5mm thick or aluminium 3mm thick. It needs to fit loosely inside the tube part of the pump body - about 0.5mm clearance all round is fine. If the piston seals well, it may pump better, but the tube is probably not perfectly round, so it will still need to be a bit loose to avoid jamming in some positions (hence the need to bore the tube if possible).

The large industrial or agricultural pumps have leather or rubber seals to make good contact with the tube wall of the pump, the lightweight nature of this wind pump means that the friction introduced by such a refinement outweighs the benefit.

The stem is attached to the piston by threading the rod M5, for about 10mm and similarly tapping the piston. Sufficient of the threaded stem should protrude through to permit a lock nut to be screwed on to the end of the stem.

The piston incorporates a valve which was covered in detail in the last section.

See Figure 35 [piston and stem]

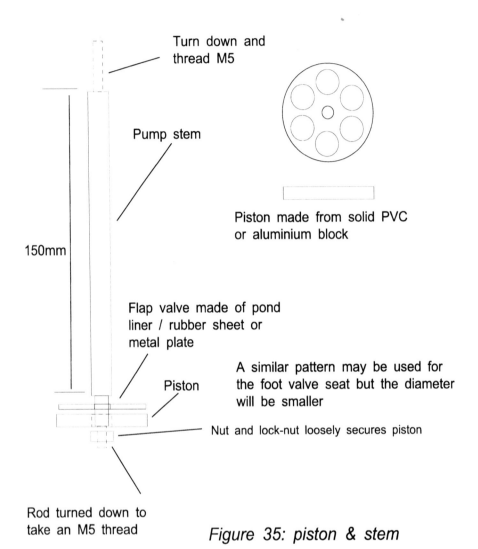

Turn down and thread M5

Pump stem

150mm

Flap valve made of pond liner / rubber sheet or metal plate

Piston

Piston made from solid PVC or aluminium block

A similar pattern may be used for the foot valve seat but the diameter will be smaller

Nut and lock-nut loosely secures piston

Rod turned down to take an M5 thread

Figure 35: piston & stem

pump filters

The filter that is fitted to the commercial foot valve mentioned above is good. So why do I not include such a device on my home-made foot valve? Well, the foot valve is not compulsory unless you are pumping fluids that may clog the valve or pump, but since it comes with the commercial valve, why not just leave it there - it will probably do no harm and may do good.

My experience has been that the home-made foot valves do not allow the pump to clog even without the filter but it may depend on what you are pumping.

plumbing

See also 'pipework and layout' on p. 94

Figure 36 [turbine tower showing a plumbing arrangement] shows a typical configuration of the plumbing from a bath tub outlet to the reservoir from which water is pumped.

It may seem that the source of the water is higher than the final destination - so why bother to pump it at all?

Well, in the original solution to the irrigation problem, I did simply drain the bath to the garden - but without using VERY expensive large-bore pipework, it took up to an hour to empty a moderate bath tub - hence the fast large-bore pipe to the reservoir and pump uphill from there.

In action, the water will rise up the stand pipe until it reaches the T-piece where it will leave the pump assembly and join the general plumbing. Assuming that there are no restrictions in this plumbing, the water will flow away with no problems. If however, the plumbing involves a long journey through fairly narrow-bore pipe, the restrictions will cause a back pressure and water will rise higher up the stand pipe.

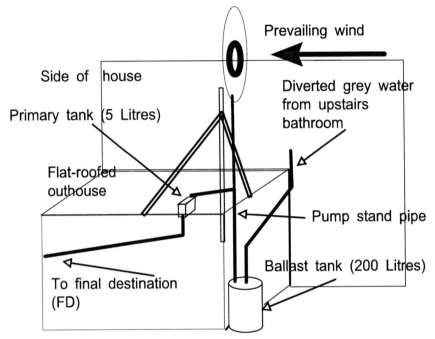

Side of house

Prevailing wind

Primary tank (5 Litres)

Diverted grey water
from upstairs
bathroom

Flat-roofed
outhouse

Pump stand pipe

To final destination
(FD)

Ballast tank (200 Litres)

*Figure 36: turbine tower showing
plumbing arrangement*

In a large wind pump the problem of back pressure is
overcome by a 'gland', also known as a stuffing box, which
seals the top of the stand pipe while permitting the movement
of the transmission rod. Such a device would introduce too
much friction on a small unit and the best way to
accommodate the back pressure is to continue the stand pipe
up above the T-piece to a distance found by trial and error.

In any event, it should be apparent that using this method,
you cannot pump water above the height of the turbine!

There will be more about specific plumbing as applied to the whole system later.

pointing into the wind

The wind does not always come from the same direction and for the majority of horizontal axis turbines, some mechanism is required to ensure that the rotor faces into the wind. There are several very good designs that manage this by rotor design alone. 'Proven' wind generators (see 'resources'), for instance, use a rear-mounted rotor which always points into wind; the only problem, if any, is that the rotor is then in the shadow of the generator; however the Proven generator has a very small diameter for its power and this therefore, is not a problem. I have not seen any rear-mounted rotors on wind pumps but they may exist or could possibly be designed.

For our turbine, the simplest method is to fix a backward-pointing pole on the back of the turbine head that has a vertical fin, like the tail of an aircraft. When the rotor is not facing the wind, a correcting force is experienced by the fin as a result of the wind, and the turbine head rotates on the vertical axis to face the wind.

Figure 23 illustrates this simple design.

furling

When the wind turbine is subject to wind speeds in excess of a certain maximum, a number of things can happen. In extreme cases the whole pumping system will be destroyed, garden sheds and trees along with it; under less stressful conditions the pump and turbine may be damaged through over-speeding. The normal solution to this problem is to brake or 'furl' the turbine rotor.

The idea of furling is to point the turbine out of the wind and thus reduce the cross-section gathering energy from the wind.

Furling mechanisms require energy to operate, and so far I have not managed to find an economical way of including a viable mechanism on these small wind turbines. If the wind looks as if it is getting too high, I stall the turbine with a rope or physically remove it from the mast; it only takes me 15-20 minutes.

This is of no interest to you if your system is some distance from you when the wind builds up; it may well be that you are not even aware of the local wind speed.

So I have examined some ideas for furling. The two main purposes of a furling device are:

1. to reduce the area of resistance to the wind, so that the wind will not destroy the turbine, tower etc.

2. to prevent the turbine from over-speeding, and damaging the turbine, gears (if present), pump and transmission system.

From this second point you can see that if all you do is simply apply a brake to the turbine, you may help avoid over-speeding, but the area facing the wind may be unchanged so damage may still occur.

I would like to focus on the first option and hope that in doing something about that, we will solve the over-speeding problem as well.

You may like to experiment with some ideas yourself; here are some results of brainstorm sessions held with some engineering pals.

One idea is that when the pump raises too much water too quickly, the excess is allowed to overflow into a leaky bucket that is attached to the fin / tail by levers; the fin and tail are pulled round by the extra weight and the turbine is pushed out of the wind. The leaky bucket empties, the fin swings back and the turbine is back to normal.

The problem with this is that it is complex, and if the initial over-speeding were to damage the pump or transmission, there would be no excess water to load the bucket.

Another idea: a secondary tail fin sticks out to the side of the turbine; this is again attached by levers to the main fin, which is spring-loaded in the 'normal' position. As the wind speed increases, the secondary fin is pushed backwards and the main fin is moved as a result, causing the turbine to move out of the main wind flow.

This is a nice idea and I believe it is already used by some designs that I have seen in Brittany; but its reliance on springs should sound alarm bells.

If the spring jams or breaks, the system is unpredictable and could end up oscillating from one extreme to the other.

A minor change to this idea does however have possibilities:

Figures 22 and 23 show the parts of the turbine head that make up the tail, fin and method of attachment for a non-furling tail fin; this will simply keep the turbine facing the wind, it will NOT furl.

What follows is the result of some experiments I have carried out to explore a possible solution to the lightweight yaw problem. Prototypes of this system have worked but haven't been tested adequately enough to come with any recommendation. However, these notes may lead you to a solution that is better and more reliable; please let me know!

The new fin assembly has a horizontal fin as well as a normal vertical fin (on the same pole).

The fin stick is on a hinge that hangs it down 45 degrees. Also, the axis of the hinge is at 45 degrees to the horizontal axis of the turbine spindle, as well as being at 45 degrees to the vertical axis.

Viewed from the front, the hinge axis looks like this: /
Also, viewed from the side, the hinge axis looks like this: /
(where / indicates a 45 degree angle).

To help visualise this arrangement, do the following:

Imagine your back is the turbine rotor, with the wind coming from straight behind you. Stand with your right arm pointing straight out in front of you (you are pointing to where the wind is going). With your right thumb on the top of your hand, rotate your arm anti clockwise through 45 degrees, now fold your right wrist so your hand is pointing down to your front left.

A rod perpendicular to the back of your right hand is now the axis of the hinge-pin on which the fin pole rotates.

From the position of your arm when you started this exercise (pointing in front of you, to where the wind is going), if you lower your arm to point down at 45 degrees, that is the direction of the fin pole at rest.

Place your left hand against your right shoulder and use your left index finger to indicate the direction of the hinge-pin. Now using your right arm as the fin pole, it should be apparent that as the fin pole rotates on the hinge-pin, it moves up and to the left until it is pointing above your left shoulder.

In this design, the fin pole has two blades: one in the vertical plane, and one in the horizontal plane (with reference to the fin pole axis).

It is a lot easier just to make one, honestly! I spent many months during which the concept developed in my mind, and once gelled, it took a couple of hours to build a test prototype. It has taken me weeks to actually draw the thing using my primitive drawing software and skills!

This configuration causes the following behaviour:

At rest, the vertical fin is vertical and slightly offset to the right, (as seen from the wind's point of view); the horizontal fin is hanging down, highest edge at the front.

In light winds the fins will lie behind and below the turbine and thus keep the turbine facing the wind, the vertical blade being primarily responsible for this control. As the wind increases, pressure on the horizontal blade increases and causes the fin to rise on the horizontal component of the hinge.

To do this, it will also be constrained by the vertical component of the hinge; thus the tail will move away from behind the turbine to the left, (viewed from the front). In doing so, the turbine will be pointed progressively away from the wind.

The front of the pole will need a counterweight or the fin will not lift and turn except in VERY high winds.

The maximum theoretical displacement is about 45 degrees, which means that the turbine rotor will never be fully furled. I have not had the opportunity to test this system in really high winds, but I think it would fail; so I repeat my earlier suggestion - if it looks like there's going to be a storm, take it down.

I have built and tested this mechanism and it does behave as predicted, but it has never been tested to destruction because I have not flown one in high enough winds (they were never around when I was testing). Also, I may have been lucky in the selection of dimensions for this design, so slight variations *may* cause unstable behaviour.

Look at Figure 37 [angled fin pole bracket]. This drawing represents the steps in preparation of the end of the fin pole bracket for the swinging fin. I have not included the part of the bracket that has the right-angle bends and attaches to the turbine head.

The swinging fin pole is shown in Figure 38 [swinging fin pole]. The hinge tube fits over the pin on the fin pole bracket and is retained by a washer and split pin - or you could make the pin from an M6 bolt and secure with a nut.

You may need to bend the fin pole down at the fin end to increase the wind drag on the horizontal blade.

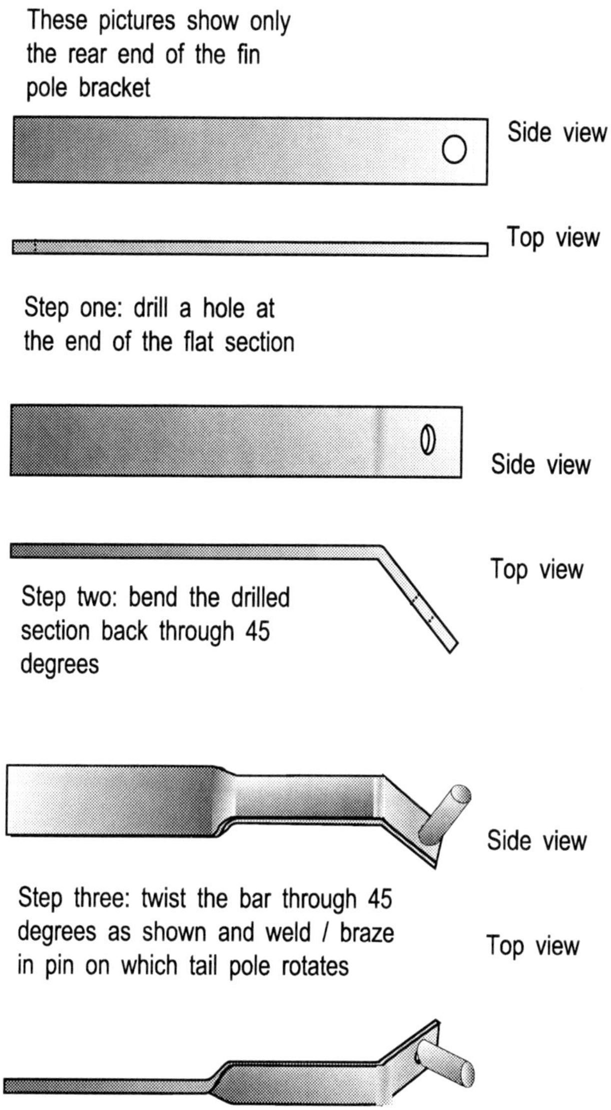

These pictures show only
the rear end of the fin
pole bracket

Side view

Top view

Step one: drill a hole at
the end of the flat section

Side view

Top view

Step two: bend the drilled
section back through 45
degrees

Side view

Step three: twist the bar through 45
degrees as shown and weld / braze
in pin on which tail pole rotates

Top view

Figure 37: angled fin pole bracket

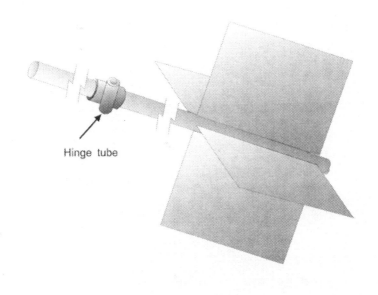

Hinge tube

Figure 38: swinging fin pole

The rest of the book relates to setting up and maintaining the system, and looking after water; but before that I need to give some detail about the final design of the turbine system. The main changes are the inclusion of a geared (chain drive) turbine head and a larger-bore pump.

If you refer back to Figure 6 [exploded view of latest (geared) main turbine components] you will see an overview of the geared head. I could simply show a design that used standard cycle components - but the variety of sizes and types of system that exist worldwide is so large that most folk will probably have something to hand that does not match my specification.

In any event, I have not recorded any dimensions on the gear-pump design, so here are a few notes on the nature of the gear-pump system to explain the layout of the components.

You will note that the actual head is, in principle, the same as all the others. However, a length of angle iron is welded to the side of the horizontal head tube, pointing down and to the left, (as viewed from behind, looking into the wind). Make sure it does not interfere with the head support tube.
This angle iron carries the large sprocket on a suitable bearing and an adjustable jockey pulley to tension the chain.

For the jockey, I simply use a piece of nylon or PTFE rod attached firmly to the angle iron, it does not turn, using the natural bearing surface of nylon or PTFE to avoid friction. If you have a suitable sprocket then use it, but mine works fine without. The block will wear a bit but it can be replaced or reversed or shimmed out to take up the wear.

An important point about the angle of the angle iron to the vertical: during the maximum load cycle of the con rod, when lifting water, the con rod needs to be as near as possible parallel to the push rod, so try to ensure that the large sprocket is aligned such that the load arc of travel is centered on that line, i.e. the line of the push rod.
In my drawing, the large sprocket is shown too far offset to the left. Had I fitted a larger sprocket (as planned), this would have been better, but as it is, the effect of this discrepancy was not in any way a problem.

Figures 39 and 40: my drawings do not give exact dimensions, but if you stick within the scale suggested by the drawings you will have a working system. It is not just the diameter of sprockets or number of teeth or even teeth per inch - it's the axial depth of the sprocket that will affect your design. See what works.

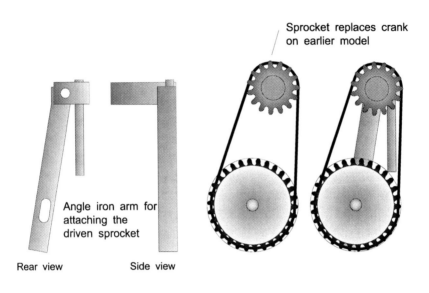

Sprocket replaces crank
on earlier model

Angle iron arm for
attaching the
driven sprocket

Rear view   Side view

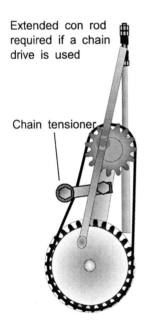

Extended con rod
required if a chain
drive is used

Chain tensioner

The con rod is attached to the
sprocket in the same way that it
would have been attached to the
crank in the older design

Figure 39: layout for a chain-driven turbine

Apart from the added components (sprockets, chain and chain arm), most of the original design is compatible with the new design. The exceptions are that the new con rod is longer and the crank to spindle fitting is replaced by an appropriate sprocket to spindle fitting.

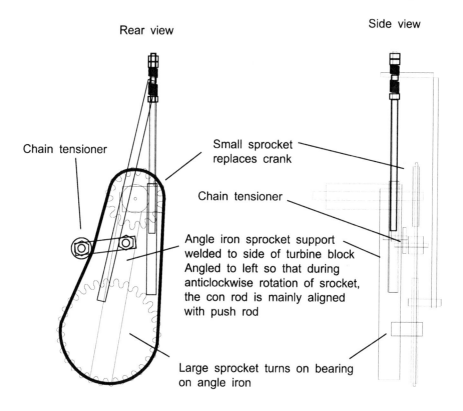

Rear view

Side view

Chain tensioner

Small sprocket replaces crank

Chain tensioner

Angle iron sprocket support welded to side of turbine block Angled to left so that during anticlockwise rotation of srocket, the con rod is mainly aligned with push rod

Large sprocket turns on bearing on angle iron

Figure 40: schematic diagram of a chain-driven turbine

This diagram shows the chain drive system as a schematic. I hope it shows more of the method of build.

new pump design

The small size of the turbine rotor means that the rotor RPM is high. The reciprocating pump and the transmission prefer to run slowly, hence the geared system. The geared system has far higher torque and permits the use of a larger-bore pump; the result should be a system that pumps a bit more water but also that lasts longer between repairs.

Figure 41 [large-bore pump for use with geared drive] shows the new, larger pump with a commercial foot valve.

Notice that the reducer at the top, from riser to bore, is female to female. However, if you use the 1.25inch 'EveryValve' commercial foot valve, you will need a female to male reducer between the pump body and valve. Also note that if you use the 2inch foot vale (see below) with the large-bore pump, you do not need the reducer at all.

This larger pump has not been tested extensively, and you may decide to use a larger foot valve, (also from EveryValve - see resources). The extra torque of this system could also cause damage to the transmission if the pump should jam, so care must be taken to build this larger pump carefully.

The design for the piston will be a larger-scale version of the smaller pump; the dimensions of the stem and attachment method are the same.

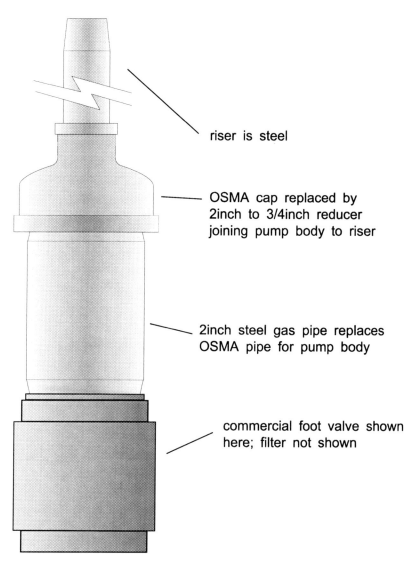

riser is steel

OSMA cap replaced by
2inch to 3/4inch reducer
joining pump body to riser

2inch steel gas pipe replaces
OSMA pipe for pump body

commercial foot valve shown
here; filter not shown

Figure 41: large-bore pump for use with geared drive

the wind pump in operation

pipework & layout

The detail of your plumbing arrangements will be peculiar to your own situation and requirements. You may need to bring grey water and rain water to a central point; or you may have one source of water only. These and other considerations will be very important to your choice of layout. Most of these choices are made using common sense, but one or two factors have affected my own final system where I have needed to store the water.

There are several reasons for storing water. They usually relate to the fact that the wind may not blow when you have water to pump. Alternatively, you may not want to move all the water when the wind does blow.

Three common reasons for needing a delay in water flow are:

1. to slow down the flow through a filter.
2. to maintain an 'ornamental trickle'.
3. to save water for later.

None of these actually describes what to me is the main reason for storing water.

A major source of domestic grey water is the bath tub and this supplies sudden surges of up to 15 gallons (c. 70 litres) that must be drained from the bath quickly, but may need to be stored for a while before there is a breeze to pump it away.

I will describe a typical scenario which will illustrate the conflicting needs of the system.

To collect grey water from all domestic outlets may involve quite complex arrangements. The example here simply looks at the bath tub.

See Figure 42 [plumbing layout].

It is quite likely that the upstairs bath outlet is higher than the final destination of the water. Here the problem is to empty the bath as quickly as possible, but still allow gravity to get the water as near to its destination as possible. Let us suppose that the bath outlet is 3 metres above and 40 metres from the final destination (FD).

The water will flow from bath to FD quite well but to make sure that the bath empties quickly enough, the plumbing needs to be about 50mm bore; this is expensive.

Instead, a tank could be placed close to the bath outlet and 1 metre below it, and connected to it via a 50mm pipe. This tank would fill fast and still drain by gravity to the FD via a normal hose pipe. But can you fit a 15-30 gallon (70-140 litre) tank in such a position safely and aesthetically? Probably not.

If, on the other hand, the tank is at ground level and is fed directly from the bath outlet via normal diameter waste pipe, the bath will still drain. Once the bath is empty and the tank is full, the pump is essential since all the head advantage of the bath to FD has been lost. Also bear in mind that if the tank overflows for any reason, the excess water needs to be directed to the drain. (See also 'water extraction and the law' - p.14.)

Let's assume that with an 'average' wind blowing, your wind pump will move a certain amount of water and that the pipework to the FD is designed and costed to allow that flow rate. On a good day, when the wind is a bit stronger than usual, or when the wind is gusting, there will be surges of flow from the pump that the plumbing will not cope with. To accommodate these surges, I have fitted a 5-20 litre primary tank, level with the top of the pump outlet. During short bursts of pump activity, this tank will fill briefly, because the pipe to the FD will not empty the tank fast enough. The 5-20 litre tank is shown on the roof of the outhouse in Figure 36 [turbine tower showing plumbing arrangement].

Notice also that the small primary tank is between the bath outlet and the ballast tank. In this position, the primary tank will still smooth out surges in the pump rate, but it can also divert the bath water to the FD, and because the flow is too much for the pipework to the FD to cope with, the excess water will overflow to the ballast tank. So even with no wind, some water will be diverted to the FD. This idea only works if the FD is lower than the bath outlet.

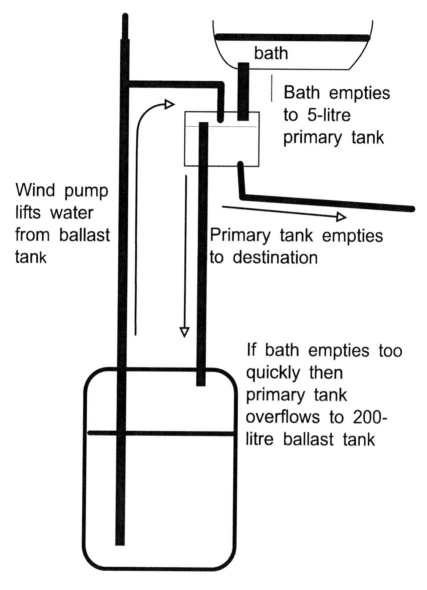

bath

Bath empties
to 5-litre
primary tank

Wind pump
lifts water
from ballast
tank

Primary tank empties
to destination

If bath empties too
quickly then
primary tank
overflows to 200-
litre ballast tank

Figure 42: plumbing layout

water storage

Preventing contamination and souring

In the last section, a requirement for water storage was identified. However, the grey water that accumulates in the ballast tank (water butt) may quickly go sour and smell if there is no wind and the pump is not working.

Water in long runs of pipe will suffer a similar fate.

If the water is going straight to irrigation, the souring of the water may not pose a problem, if however the water is ultimately going to a pond then some filtering and aeration may be required.

It is becoming more popular in the UK to recycle grey water to create a non-potable secondary supply, usually just to flush toilets. This recycled water cannot be used for washing, let alone cooking, and it must be VERY well filtered before it is used even to flush toilets.

Failure to do so will result in foul smells from the cisterns, frequent failures of the valves in the cistern and general blockages in the pipes.

Passing the water through a bed of shingle in which reeds are growing does take out quite a lot of the visible and aromatic contamination, but this slows down the flow. If the contamination is not too severe, a reed bed 1 metre square for every member of the household should be adequate; but don't expect the water to be always perfect after such a small filter.

If you have space for 5 square metres of reed bed per person, you can actually recycle foul water but you need to re-circulate the water through a 'pond' many times before it is clean. This is not my subject, but there are companies who specialise in this sort of thing (see 'resources').

You need to experiment to establish the optimum flow rate for the system, to minimise the deterioration of the water being stored, while at the same time slowing the flow down to maximise cleaning by any filter system.

If the ballast tank is too large, the water will stagnate; on the other hand, a large ballast tank will have the effect of diluting the occasional large single dose of bathroom cleaner which is accidentally routed to the fish pond via the pumping system.

filter to remove solids

Some form of mesh filter will help to remove solids, but it needs to be very large - say 500mm diameter. The reason for this is that the solids will be small and in low volumes - it is, after all, mainly bath water. The problem is that after a short time a healthy (?) layer of living organic matter will grow in the filter. If this is not a nice growth, then you need to clean it regularly - if only for hygiene. But even if you are lucky and the growth is not unpleasant, it will still block the filter or at least reduce the flow.

I need to distinguish here between the organic filter used to *clean* water and the filter in the pump system that prevents the pump from clogging. If you are pumping fluids that contain a great deal of suspended matter then the pump filter will need to be cleaned regularly. This is not a problem if you are simply recycling domestic grey water as the pump is probably accessible.

organic shingle filter

The large filter beds associated with commercial sewage processing plants rely on a continuous trickle of mechanically-filtered water permeating through a bed composed of sand, shingle and ballast. The bed is always moist and acts as a medium for the growth of aerobic micro-organisms, which will devour the unwanted organic impurities in the sewage.

The filter bed must not dry out, nor must it become waterlogged for too long, as either scenario will kill the micro-organisms.

In this type of system, the sewage is delivered to the filter bed by means of a series of sprinklers that slowly rotate above the circular bed, powered by an electric motor.

This type of system does not scale down very well and has the added disadvantage for the user of renewable energy that it relies on electric power. Having said that I have visited one site where a community of 40 people rely on just such a system that has no electric power and a 7-8 metre diameter rotary sprinkler filter system that has been running for over fifty years. The only maintenance is that the surface of the gravel is raked once every month. The whole system is a bit smaller than a tennis court, so it is not really suited to the average family home.

The system I have developed is small and emulates the large-scale filter quite well. Instead of allowing water to continuously trickle through a bed of shingle, I allow a tank of shingle to fill with grey water and then empty completely by means of an automatic syphon. In this way, the shingle, and of course the associated micro-organisms are alternately immersed in water and then allowed access to the air.

I mentioned earlier that the filter must not be waterlogged or dry for too long; bearing in mind the likely source of the water in a domestic situation, there is a third danger of non-organic, non-biodegradable, and possibly toxic cleaning fluids ending up in the system.

Small amounts of soap and detergent are OK, but bleach and washing machine or dish washer powder must be avoided if possible. I recommend only using environmentally-friendly soaps, washing powders, washing-up liquids etc, and not to use bleach at all.

Some more enterprising folk who wished to recycle water from dishwasher and washing machine have used an optical detector in the outlet line so that water is diverted straight to the drain unless it is fairly clear, i.e. the system is in rinse cycle.

Back to the filter: a suitable tank for a filter is the type normally used as a central heating header tank in a domestic loft.

Figure 43 [design for a tidal biological filter] shows the plumbing.

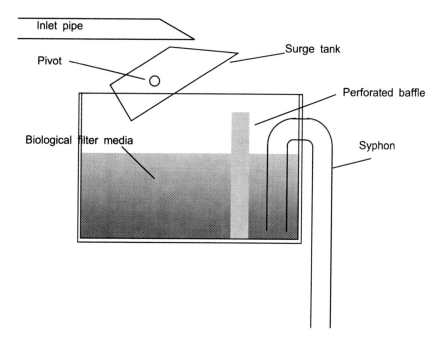

Figure 43: design for a tidal biological filter

Automatic syphons will not work unless the feed rate exceeds a certain minimum, determined primarily by the bore of the syphon tube. To ensure that there is a good surge of flow to the tank, even when the feed is only a trickle, I use a 'nodding dog' - so called because when I first started building them, I made them from a large dog food tin (and it nods).

I am told that a similar device, made from bamboo, is used in China as a bird scarer.

The operational principle of the nodding dog is that the tin can is stable in an upright position when empty, but as it fills with water, the centre of gravity drops initially, but rises as the water level rises in the can. If the pivot point of the tin can is correct then the can will continue to fill until it becomes unstable and falls over, releasing its contents. With its contents gone, the can will return to its stable upright position, ready to fill again.

Each nod of the dog gives a 1-litre surge of water to the filter tank. When the level of water is near the top of the syphon tube, the next surge will start the syphon.

The syphon must be capable of emptying the tank faster than the incoming water can fill it. This is usually no problem, but take care to monitor it on windy days as well as not-so-windy days.

Experiment with different sizes of tank and nodding dog.

I use a typical central heating header tank, about 300mm * 200mm * 250mm. I now use a 2-3 litre plastic lemonade bottle for the nodding dog; maybe I should call it a nodding lemon.

With these components, it helps if you can make the lower end of the syphon tube at least 250mm below the surface of the filter tank.

maintenance

A commercial wind pump would be expected to run for weeks on end with no attention at all, but in some parts of the world, where for instance cattle are depending on the water supplied by a series of wind pumps, the stockman cannot risk water supply drying up, and so the pumps on a cattle station will be visited every few days, just to check all is well. There may be several pumps spread over many hundreds of square miles and the stockman may take 2 or 3 days to visit all the pumps in one round trip, so the job is a bit like painting the Forth Bridge! The value of the wind pump is small compared to the value of the stock that would die if the pump failed for more than a couple of days.

The small wind pump described in this book is not as robust or reliable as the commercial ones described above but the circumstances of its use are also very different. You are probably going to see the system most days and will hear any odd noises that it makes. If it fails, the cost will be measured more by what it falls on or the cost of repair of the wind pump itself.

The parts that suffer from wear are the main turbine spindle bush, the ends of the con rod, push rod and crank, and the push rod sliding in the head vertical tube. All these parts should be fairly quiet apart from a gentle 'swoosh' and 'click'; any squeaks usually mean that some oil or grease is needed. Take care to avoid letting oil reach the pump and the pumped water, especially if the water is going to a fish pond etc.

This sort of attention should be given every few weeks or when needed. Whenever the system is being examined for any reason, always check the structure of the whole machine - this means the rivets in the turbine, the tail fin, the brazing of the head and all connections and screws etc.

Do *not* paint any parts of the turbine assembly. Painting simply covers up any tell-tale cracks in the structure.

If however, you have access to galvanizing facilities, use them on any materials that can be protected in this manner.

In a domestic environment, the pump is usually more accessible than the turbine, but bits will wear out and fail. You can engineer the mechanism far more robustly than I have and this may pay off, but at a cost.

My attitude is to keep an eye on the pump, and if it breaks, fix it - otherwise leave it alone. An exception to the 'wait until it breaks' attitude arises if you notice a gradual drop in performance; it may then be worth stripping the pump down to look for leaking valves or cracks in the body etc.

resources

materials

pump components

The OSMA range mentioned in the book can be purchased from:

- Wavin Plastics Ltd
  Parsonage Way
  Chippenham
  Wiltshire
  SN15 5PN

  +44 (0)1249 766600

  www.wavin.co.uk

you can use the search facility on their website to find components using the code numbers used in this book

nutserts

I mention these early on as a method of making a short-lived bearing in an aluminium con rod. These handy fixtures are like a normal nut except that the normally hexagonal shape at one end is turned down to a smaller diameter circle. In use, this round section is inserted through a suitable hole in sheet metal and the round section is expanded outwards to engage with the hole in the sheet. The process of expansion is usually done by squeezing an oversize ball bearing into the bore of the round section. A suitable nutsert for the purpose described is an M6 plated mild steel. They may be bought from:

- Alfast Ltd,
  2 Gloucester Road,
  Luton,

Bedfordshire LU1 3XH
UK

+44 (0)1582 418498

http://www.alfast.co.uk

small-bore rigid PVC tube

The tube mentioned in the text for use as transmission rod and transmission rod sleeve used to be obtained from:

- Wilford (Plastics) Ltd,
  Cosgrove Way,
  Luton, Beds

  +44 (0)1582 36961

I am sorry to say that they are no longer there. I have failed to locate a new supplier, and although I have a large stock of this material, I am not able to supply it myself - but if you got this far you will probably find an alternative.

scaffold tower

The 8-metre length of scaffold pole referred to in the text needs to be made by joining two shorter lengths (the maximum standard length is about 17 feet, or around 5.2 metres). If you have access to this type of material, it is a very convenient method of construction.

foot valve

For the commercial foot valve mentioned in the text contact

- EveryValve Equipment Limited,
  19 Station Close,
  Potters Bar,
  Hertfordshire, EN6 1TL

UK

+44 (0)1707 642018

sales@everyvalve.com

www.everyvalve.com

They are a very friendly company who will deal with your requests efficiently.

salvage yards

- www.salvo.co.uk - a list of salvage yards in Britain and internationally; salvage yards are excellent sources of recycled building materials. This site will help you find your local yard.

commercial manufacturers of wind pumps

- Poldaw Windpumps
  Neale Consulting Engineers Ltd,
  Highfield, Pilcot Hill
  Dogmersfield,
  Hants, RG27 8SX

  +44 (0)1252 629199

  www.tribology.co.uk/poldaw.htm

wind pump information

- Reading this book may prompt some questions; I recommend a book for some answers: Peter Fraenkel, Roy Barlow, Frances Crick, Anthony Derrick and Varis Bokalders, *Wind pumps A guide for*

development workers, Intermediate Technology Publications in association with the Stockholm Environment Institute, 1993. ISBN 1 85339 126 3
- http://www.itdg.org/docs/technical_information_service /windpumps.pdf - technical briefing on wind pumps from the Intermediate Technology Group
- http://www.bwea.com/ref/pumps.html - info on wind pumps on the British Wind Energy Association website
- http://www.gamos.demon.co.uk/just%20gamos%20ho mepage/henkfnl2.htm - interesting 'rope' wind pump design for developing countries
- http://igadrhep.energyprojects.net/Links/Profiles/WindP umps/WindPumps.htm - wind pumps and renewables for Africa; includes some technical info and a glossary

greywater

- LILI run a residential weekend course called 'Sustainable Water & Sewage', which covers greywater recycling, reed beds, trench arches, rainwater collection, water saving and compost toilets. See back cover for contact details
- Art Ludwig, *Create an Oasis with Greywater*, 1994, Oasis Design. Wonderful manual on all aspects of greywater use
- Art Ludwig, *Builder's Greywater Guide*, 1995, Oasis Design. Follow-up to the above manual, with more technical information on installation
- www.oasisdesign.net - great resource on greywater and other environmental technologies

water & plumbing

- www.fwr.org/nwdmc.htm - Environment Agency Water Demand Management Centre; from this site you can order free copies of their factsheets on a whole range of ways to save water, from compost toilets and

rainwater collection, to water-saving appliances and greywater recycling
- www.wras.co.uk - water regulations for the UK. includes an order form for the complete regulations - costs £16.30 plus p&p
- R D Treloar, 2000, *Plumbing, Heating and Gas Installations*, Blackwell – the best plumbing book there is
- Stu Campbell, *The Home Water Supply*, 1983, Garden Way Publishing. Down-to-earth solutions for all aspects of domestic water, for plumbers and non-plumbers alike

rainwater harvesting

- There is a free factsheet on rainwater harvesting on LILI's website
- Klaus Konig, 2001, *The Rainwater Technology Handbook*, from Green Shop 01452 770629 – best book on the subject
- Arnold Pacey and Adrian Cullis, *Rainwater Harvesting: the collection of rainfall and runoff in rural areas*, 1986, Intermediate Technology. Focuses on small communities in developing countries
- *The Worth of Water:* technical briefs on health, water and sanitation, Intermediate Technology. Essays on developing country village and community-level water and sanitation
- Environment Agency - 01903 832073 – contact them for *Harvesting Rainwater Information Guide* and lists of suppliers
- www.rainharvesting.co.uk 01452 772000 – suppliers / installers plus information on their website
- www.thetankexchange.com - 08704 670706 - recycled industrial fruit juice barrels
- Plantpak (rainsava) - for diverting water from downpipes to a tank or water butt. 01621 745500

http://www.desch-plantpak.co.uk/ishop/488/shopscr35.html

eco soaps and cleaning products

- There are many health and environmental problems associated with synthetic chemicals when they end up in our bodies or down the drain (or in your greywater to be pumped to your garden etc). Over 70,000 synthetic chemicals are manufactured, and only 600 have been adequately tested (US Office of Environmental Affairs).
- There is a free factsheet about natural cleaners on LILI's website, with natural recipes for a range of cleaning products
- *Hazardless Home Handbook*: lists most household products in alphabetical order, problems associated with them, storage, disposal, and natural alternatives; download it free from LILI's website - www.lowimpact.org/booksnaturalcleaners.htm
- *The Good Shopping Guide*: 0845 458 9911; www.thegoodshoppingguide.co.uk - ranks brands using a range of criteria - anything you could possibly want to buy
- The Green Shop – www.greenshop.co.uk - for green cleaning products; they have everything you need
- http://www.nancysnatural.info/recipesandtips.html - fun green cleaning recipes and tips

reed beds

- Elemental Solutions
  Withy Cottage
  Little Hill
  Orcop
  Herefordshire
  HR2 8SE

  +44 (0)1981 540728

www.elementalsolutions.co.uk

wind power

- British Wind Energy Association - trade Association for wind energy in the UK

  BWEA
  Renewable Energy House
  1 Aztec Row
  Berner's Road
  London N1 0PW

  +44 (0)207 689 1960

  www.bwea.com

- Proven - wind turbine manufacturers

  Proven Energy Ltd
  Wardhead Park
  Stewarton
  Ayrshire
  KA3 5LH

  +44 (0)1560 485570

  www.provenenergy.com

notes